Alone on the Mountain

Alone on the Mountain

SHEEPHERDING in the AMERICAN WEST

PATTI SHERLOCK

Illustrated by Earl Thollander

Doubleday & Company, Inc., Garden City, New York
1979

Library of Congress Cataloging in Publication Data

Sherlock, Patti.
Alone on the mountain.

1. Shepherds—The West. 2. Sheep—The West.
3. Shepherds—Idaho. 4. Sheep—Idaho. I. Title.
SF375.4.W4S53 636.3'08'30978
ISBN: 0-385-15098-9
Library of Congress Catalog Card Number 78–7768

Printed in the United States of America
First Edition

Alone on the Mountain

I

Introduction

The sheepherder sits on an Idaho mountain and watches his sheep in the valley below. The sheep are his responsibility, his work, his amusement, sometimes even his friends.

To thrive, the sheep need peaceful conditions free of annoyances, safe from predators. Their grass must be abundant, water adequate and rest undisturbed.

The flock's welfare comes before the herder's comfort, concerns and hungers, whether for food or other human company. His callused feet and torn legs tell of a lost lamb that ended in a thistle patch. His sagging eyes tell of a sleepless night watching for the coyote that has raided his flock and killed several lambs. His fingers whittle human features onto a pine branch to salve his loneliness.

Vertical patterns carved by the wind crease his narrow cheeks and squint lines cut by the sun bury his eyes. But a lean, straight body mocks the prematurely old face. Heroic western distances have strengthened his lungs and legs. When he senses a sheep is in trouble, he springs over rocky terrain to track it.

A sheep shack, a metal-covered wagon, is the herder's home. Its interior is spare, its location lavish. Blue-gray sagebrush, orange paintbrush, towering pines and tiny bluebells decorate the hills where the herder walks. Ranges of snow-capped peaks lie below his lookout points.

The sheep in his charge, about a thousand ewes and lambs, are not a monotonous bunch. Each has a distinct face and private behavior. The old ewe with yellowed fleece has watery eyes that recount arduous trails and lost offspring. The Suffolks, with wise black faces, teach their annoying, straying ways to their lambs. The pesty lamb with expressive face trips the herder in its efforts for attention.

When the sheep wander too far out the herder motions for the dogs to herd them back. He tries not to "overdog" the flock and he insists that the dogs work quietly. The dogs, too, develop an enormous loyalty to the flock and will stay at their posts, ignoring their own weariness.

At night he feeds his dogs and his horse, and the sheep bed down in the meadow. He walks among them to check their condition. If they look well fed and healthy and if the location is safe, he crawls into bed in his sheep shack.

In early morning's whiteness before the sun appears, he hauls himself up. He saddles the horse, feeds the dogs and gets the sheep moving in the cool. If they are trailing they must cover several miles; if they are stationary on summer range he will move them to fresh grass to protect pastures.

If you happen by his camp sometimes, sit with him awhile. He will be glad for your visit. Probably he won't talk much. Perhaps your languages are not the same, but even if they are it is good to economize words on the mountain.

The scene may call into your mind a poem, "The Lord is my shepherd, I shall not want." The herder knows David's song. It strums the cadence of the sheep cycle and taps out the meter of the changing seasons. His life flows to its music.

He understands about God's love and care for His flock. Everything else has changed since the shepherd boy David wrote his song. But on a modern mountain a man can still know and understand the same things David did.

But what he knows and what he has learned he can't pass along to you. Even if he had the vocabulary he'd be too shy to use it; years of solitude have made him uncomfortable with people.

You would have to stay up on the hill with him. You could watch him tend the sheep and you could learn something, what David knew about God's ways with men, and about how much a good shepherd loves his flock.

II

The Lord Is My Shepherd

In the shadow of a purple mountain, a lamb is born.

On its birthday, corn snow batters the drop herd, a group of expectant ewes penned behind worn wooden fences in an amber dip between two foothills.

The sky is low; it appears closer than tall, snow-coated mountains only ten miles away. Nearby foothills show brown earth and burlap clumps of sagebrush, and new snow sticks only to the waving weeds. The calendar says spring is near, but no hopeful indications of green are yet on the landscape.

The ewes huddle in conventions with woolly hindsides to the wind. In kinder weather at this time of day they might lie down to rest and ruminate, but in the blizzard they bunch for warmth.

Presently a young ewe with swollen sides ambles away to find solitude. She is about to lamb.

Several yards from the others, the yearling ewe paces in agitated circles. Every few steps she stops and lifts her lips to sound a complaining bleat. Her head snaps in anger at the pain while her sides heave and white woolen belly rolls.

Finally she arches her neck and kneels on her front legs. Then she slumps to her side beside a mud mound to shield herself from the hostile wind.

Because it is a first birth, the lamb's trip from womb to world is slightly long, slightly difficult.

The ewe points her nose skyward and pants. Abruptly she jumps up, sniffs and licks the earth where she has lain—then, straining, collapses back on her side.

A blood- and slime-soaked head along with two cloven feet appear beneath the ewe's tail. The birth is normal. In some cases legs tucked under the lamb in the birth passage can halt progress until a herder reaches in and cautiously guides each hoof forward. About one per cent of lambs come breach. It is tricky to turn them and often both ewe and lamb are lost.

The newborn drops to the frozen earth in a bewildered tangle of its own legs. With weak, unseeing eyes he squints up at kaleidoscopic snow swirls while the raw wind whips his wet face. His frail neck lolls. His ears flop against his naked face. When he struggles to stand he becomes a wisp of orange wool against the brown earth. He emits a helpless, human wahhh.

The ewe scrambles to her feet and stares at the new creature she suspects has some relationship to her. When she moves in for closer inspection her body interrupts the wind's cold breath, and the hopeful lamb, who knows his need better than his mother, bleats another plea.

The ewe jumps back. From a spot several feet away she gapes at the crying creature. Furtively she runs in and licks its wet, warm skin, then retreats again. The next time she runs in she rams the lamb with her head, throwing him from a collected resting position into an elongated sprawl. He just regains his upright position when his mother comes in to slam him from the other side. His reactive bleat is louder, bewildered, but with patient struggle he hauls up his wavering back legs and begins to inch on kneeling front legs to his mother. She walks away.

The shivering lamb demands warmth and nourishment if he is to survive. His wail now is weak, sad. But someone hears, and cares.

The Basque sheepherder, almost concealed beneath `hat,

neck scarf, sweat shirt, coveralls, jacket and nylon lambing chaps, is working his way through the drop herd looking for newborns.

Now his eye falls on the yearling and her lamb. On a nice day he would allow the pair more time to get acquainted—often five minutes helps—but the hostile elements make it imperative that he get the cold lamb indoors right away.

He hauls up the trembling lamb by a hind leg and starts to move toward a long, canvas-covered shed. The mother does not readily come along, but runs in a confused zigzag many yards behind. The herder backs up and pushes the lamb toward its mother's nose. She runs ovals around the herder, sniffs the thin animal dangling in his grasp, then starts to follow.

Often herders pick up new lambs in a "granny cart," a small tractor with stock trailer which holds three lambs in the manger while ewes ride behind. But because the herder is close to the shed, he decides to walk. Twice he stops, leaves the lamb on the ground and walks away, to rekindle the ewe's interest.

In future years this ewe will be more protective. Sheepmen know that ewes can leap as high as a pickup's bed to be with their lambs, and practiced mothers hover over their young.

A big ewe whose coat hangs in thick ropes sidles up to the herder. She is a granny, a dry ewe who wants the newborn lamb and will try to drive its real mother away.

When the granny menacingly shoves her nose at the mother, the young ewe pivots on her hocks and retreats. Herder, granny and newborn pass through an open gate into the shed. The barrel-chested Basque glowers at the granny.

"*Alde!*" he yells in Euskera, the Basque language. "Go away!"

The granny has complicated the herder's task, and he must trudge into the blizzard again, lamb in hand, to try to coax its mother back. The granny marches stubbornly between the lamb and its uncertain mother. Out of patience, the herder thumps the granny with his knee and drives her away.

With the granny gone, the real mother slips into the shed to join her lamb. The storm slaps the shed's canvas roof against wooden rafters and the ewe startles. Suspiciously she eyes the aisle that leads to straw-filled stalls.

The herder plunks the tiny lamb into a clean, dry stall. When he sees its mother does not plan to follow, he grabs her hind leg and drags her into the pen.

The shed is warm from the heat of 150 animals, and friendly from their blether. The herder lifts the lamb and touches a swab of tincture of iodine to its navel to prevent infection. He watches the ewe.

On wavering legs the lamb tries to find its mother's teats, but each time his nose touches her the ewe moves to another corner. Once when by luck the lamb clamps his mouth to the nipple, the ewe strikes out with a hind leg and sends the lamb tumbling.

From a nail by the shed's door the herder snatches a piece of baling twine and leaps back into the pen. The hairy rope twists in his hands, weaves through itself and becomes an instant halter. He fits it onto the struggling ewe's nose, then secures the other end to a board.

Holding the lamb beneath his arm, he squats down and finds the ewe's engorged bag. With one burly arm he steadies the ewe's rear end, with the other he guides the lamb to badly needed milk.

Even a few swallows make a difference. Almost immediately flesh around the lamb's pitiful ribs balloons and the lamb's long tail, not yet docked, wiggles gaily.

The herder smiles as he watches the lamb's expressive rear end. It is good to see the lamb do well. There is grief for the lost ones, about one in ten.

Rotating his head from left to right, the herder paces down the corridor of stalls to inspect other new lambs.

Curled in one stall's corner is a fuzzy black lamb with white head. The ewe nuzzles it proudly. Only hours before she lost her lamb, but the herder successfully fooled her and replaced her lamb with an orphan. When her own lamb was only a few minutes dead, the herder fashioned a one-piece sweater from its skin. His incision cut around the dead animal's neck and down its legs, producing a jacket with familiar scent for an orphan. Usually the ancient trick works and the engorged ewe has a lamb to suckle and raise.

Herders also try to pair weak twins or triplets with good

milking mothers of singles by rubbing slime from the ewe's afterbirth on the foster lamb. Only sometimes does this work.

For lambs who lose their mothers and no substitute is available, herders sometimes provide bottle feedings until the "bums," or orphans, can be given away or sold, often to people with acreage and children.

The herder fishes in his pocket and brings out a packet of pills. He stops at each stall and pops one in each new lamb's mouth, then rubs its throat to make sure it swallows. Most outfits now routinely gave Aureomycin to all lambs after they have suckled at least once, to stave off scours. If unchecked the virus can run through a band and claim a costly share of newborns.

The herder stops at the stall of the newest arrival. He decides to leave the nervous ewe tied, and attaches a scrap of material to her stall to remind himself to check later on this pair. In the next few days he will deal with this lamb again, to attach a rubber ring to its tail and another to its testicles, a modern, bloodless way of docking and castrating.

The shed begins to darken with the fading day. Outside other herders drive ewes who will lamb during the night into a smaller, floodlit pen where newborns can be found readily.

The new lamb's nose is tucked against its belly and its eyes are closed. The herder smiles. But he can't linger. He has more lambs to gather.

Artists have an enduring fascination with hillside flocks and the men who tend them. Inventors of children's tales and fables have spotlighted herders and their efforts to shield their animals from harm. Herders occupy a prominent berth in religious literature. Jesus repeatedly illustrated His lessons by comparing the herder-sheep relationship with that of God and man.

But these hard-drinking crusty men fit no one's romantic mold as heroes for a religious book. The book's topic intrigues me and I look forward to the research because it will be outdoors, but the herders I am meeting seem unlikely heroes, not at all the curly-headed, pink-skinned shepherds of illustrated Bible stories.

Their appearance is rough, their language and habits rougher. I've visited a couple of outfits during lambing and haven't yet found in the crews a boyish face with ethereal expression. Instead, unshaved, unsmiling faces peek out from hooded sweat shirts and cigarette-smoke haze. Purple iodine stains their jackets and blood splotches their hands and faces.

The setting where they work isn't pastoral. My leg bumps a ewe's bloated body—she died giving birth. Tiny dead newborns scattered around, sometimes gnawed by dogs, are a woeful sight. Grim weather beats the men as they do their work and, understandably, they aren't cheerful.

But were the parched hills of Palestine more poetic? To help flocks thrive on the sun-baked wastelands near Bethlehem where David watched his father's flocks required enormous toil. Near those same hills, angels first brought the news of Jesus' birth to shepherds. Were they a refined bunch? Christmas cards don't portray them with weathered faces, soiled garb and brawny muscles, but that must have been a natural consequence of their occupation. The pastel colors artists use for rural pageants tempt polite urban culture to forget the real scene's vivid earthiness.

Something else bothers me. My tendency is to idealize the herder's relationship with the flock, but I can observe that not every man has a dedicated concern for the sheep. It's a disappointment to watch a man's indifference to casualties in the herd, or kicking and cussing when a ewe is having a difficult time birthing.

I once recorded a sheepherder's funeral service. The dead herder's name was John Reed. At the funeral, John's employer Allan Thompson spoke the following eulogy:

> We are gathered here today to commemorate the life of one of our friends who has been called home.
>
> John's early life was somewhat of a closed book and although I got to know John pretty well he never seemed to have much to say about it. His name was actually Alphonso, and though very few people knew it, he hated it. My association with John started about twenty years ago, so I will have to go from there.

John was afflicted with many of the faults that most of us are, but he wasn't a hypocrite and I don't think he would want me to whitewash him here at these final rites.

I went through the Ten Commandments and I can't think of any that John didn't only break, he bent them double and stapled them shut. Yet John was a man with a lot of good qualities and he had quite a lot of principle about him.

John started working for me in the days of loose haystacks and it took a pretty good man to haul enough hay to feed the sheep. I remember about that time some of the farmers were starting to bale their hay so I would have to send part of the wagons for baled hay and part for loose. Needless to say, the baled-hay men only had to work about one fourth as hard as the others, so it created a lot of dissatisfaction, but John would volunteer to go after the loose hay and he never complained.

He took a lot of pride in seeing that the livestock in his care were well fed and cared to and he didn't hesitate to chastise the other men if they weren't feeding enough. He loved good horses and always took extra good care of the team or saddle horse entrusted to him. I had to use five or six teams at that time and I have seen John many times out in the wagon long before breakfast or after supper giving the other teams a little extra grain or picking them out some good hay.

He was one of the hardest-working men I ever knew and that is a quality that we have to admire more and more.

I remember John around the cookhouse and in his camp. He was always neat and clean and well mannered. He was courteous to the cook and other men.

I have been on the sheep trail with John with only the horses for transportation. We always have to stay with the sheep on the particular trail we go over. I remember quite a few men I would find an excuse to not sleep with or even eat with but I never hesitated to crawl in bed with John.

I think all of us here are aware that John was a hard and heavy drinker, and like most drinking men would usually go broke after four or five days in town. I have let him have money many times and he always made a special effort to pay me at the first opportunity. This is also a little unusual in this age of changing values.

I was well acquainted with John's son Daryl. He was my neighbor south of me and built a house right on a sandhill. He built this house with his own two hands and it was quite a masterpiece. I mentioned the sandhill because Daryl tamed it with no equipment but a shovel and a wheelbarrow. I know he must have been taught to work by his father and I know that a lot of his other good qualities must have come from some early training and example by his father.

Along with this I would like to mention that John showed a lot of kindness and consideration for my own father, who was close to eighty years old and in declining health. His eyesight was failing but he was determined to help with the lambing and sometimes he would miss the gates. Even I was kind of hard-boiled about it when I was so busy, but many times John would go open a gate for him.

I have described a man that you have to admire and I know that I am a little better man for having known John. I pray that I have said something that will make us all understand one another a little better and appreciate the life of John Reed.

There are still good shepherds, then—dedicated, hard-working men, no matter how rough their manners are. And no man, even if he is unmoved to begin with, can steel himself forever against this kind of natural beauty. The beauty and the solitude too must impart precious lessons.

In any case, I may need to relax my ideas about heroes for religious books. The Bible dealt with both the sacred and the profane. Jesus, in a cave, drank his first breath amidst pungent animal odors and steam from furry bodies. These herders may have much to teach me.

III

I Shall Not Want

The sheep-camp tenant lives without conveniences the rest of America considers necessities—running water, electricity and plumbing. But the herder likes his simple life.

"When I stay in town," one said, "there is a honk here, a siren there, a light here, a noise there. Up here I can sleep."

Silver-domed sheep camps, once a common part of western scenery, are scarce since the sheep industry's decline. Sheep numbers declined by three fourths in thirty years, from 52 million head in the 1940s, to 11 million in 1977. Poor prices, predators, disease, increased costs and switchover to farming and cattle all contributed to the shrink. Though the trend is turning around and stockmen can't meet present lamb demand, changes in American life-style, plus heavier demands on western land, mean sheep numbers never again will be what they once were.

Still, in rough mountain and desert terrain where sheep graze empty stretches of public and private land, nothing can replace the durable sheep camps.

Men who choose sheep camps for homes come in several nationalities and represent a variety of backgrounds, habits and refinements. In Idaho many ranchers like Basques, and Basques have won the spotlight in herder lore. But Anglo-type Americans make up a big share of herders. Mexican-American herders are numerous. American Indians are currently popular with sheepmen. So are Peruvian and Mexican aliens. At one time Scots, Irish, French and Greek immigrants all had a turn at herding.

Wages range from $350 to $500 a month, groceries and sheep camp furnished. Men find they come out better tending sheep in isolated places than when they work for higher pay in town. Money slips away in city stores, rooming houses and taverns.

Most herders lack much formal education. But because time is available, books are a standard hobby.

Every year the story goes around that an outfit down the road has a former college professor working as a herder. He wanted to escape frantic city life, the story goes, but the yarn's popularity exaggerates how often this really happens.

Some couples live out in the camps; rarer still are couples with children. Some married herders swear that life would be perfect if they could persuade their wives to join them in the hills. But lack of conveniences and social contacts put many women off. Some men return to herding after marriages fail, and many others, deprived of social opportunities, don't marry at all.

Ranchers worry who will replace the knowledgeable men in their seventies and eighties who still herd. Young men are not taking up herding, ranchers say. Generations raised with TV and telephone don't seem to have inner resources to handle the job's isolation.

Ranchers need a full crew when lambing begins, in January or as late as May, depending on methods and area. Work is endless when lambs start to come, and bosses park sheep camps near the lambing sheds. Some bosses still hire a cook and set up cook shacks where men can catch meals.

One rancher remembers a Basque herder so faithful he would not leave the sheds for a meal break if lambs were coming fast. Instead he would grab a tin, milk a ewe, down his drink and go on working.

In early spring, when trailing begins, trucks pull the camps out and park them near grazing areas. Horse teams used to pull camps, but four-wheel-drive pickups, which can go over the same hills and into gullies where roads don't exist, replaced them. Some sentimental ranchers continued to use their efficient teams until recent years when it became a problem to find men who knew how to harness and care for draft horses.

Camp tenders try to put camps on ridges where cross breezes will help cool them in midday heat. Wind helps fan insects away too.

Some herders are fussy about the camp's placement. One man used to keep a whiskey bottle half full of water to use as a level on moving day. After the camp tender moved his camp he would place the bottle on the stove, shelf and table. If it didn't show level on every test the camp tender had to try again. Dirt flew while the herder dug and leveled for his home, and often it took the camp tender several tries to put it where it suited.

A cast-iron wood-burning stove sits by the camp's entranceway. When a cooking fire burns in it the camp stays warm, but the temperature plunges when the fire dies. Spring storms can lash through the camp's thin walls and send the occupant under piles of bedding.

A bed large enough to sleep two runs the camp's width. Below it, a long bench conceals storage space for grip and groceries. A wooden square on a side wall lets down to make a kitchen table or writing desk. Hooked back up, the board hides a salt and pepper cupboard.

Behind the stove are cupboards and shelves for pots, pans and more groceries. A kerosene lamp, alarm clock, insect spray, radio and a Louis Lamour Western sit on a high shelf beside the bed. A cover girl smiles down from one wall and women clipped from horse magazines pose in western wear on another.

The camp tender comes every three or four days to deliver groceries. He'll bring water too, if a good supply isn't close. If a spring is within riding distance the herder fetches his own supply in a five-gallon canteen which he hooks over the saddle horn with a leather thong.

In the main, camps are clean, though a bachelor herder's housekeeping might not meet standards of suburbia. Hot water, no soap, poured over plates and cups passes for doing the dishes. Many men keep a grease-congealed frying pan in one corner and never wash it. The menu is flavored by meals which preceded it. Usually floor and benches are clear and the kitchen area is tidy. When sheep are grazing federal land, regulations require herders to be careful about litter and cans outdoors.

On laundry day the herder heats water on his stove and dumps his clothes in a tub. After washing them he squeezes out the water and lays the articles over sagebrush plants to dry.

If a herder wants a bath, he sets a tub of water in the sun and by late afternoon it is pleasant bath temperature. Some herders prefer to bathe in beaver ponds, finding small pools dependably comfortable.

Near the Brockman Guard Station in southeastern Idaho where the Forest Service counts sheep before they go on summer range, a natural hot pool used to attract many herders.

"They'd be lined up, ten or twelve herders, waiting for a bath," Ab Waters, a long-time herder recalls. "There was a bathhouse there and a big tub and a flume to control the water flow. But a big flood years back washed away the bathhouse and the hot pool isn't as good as it used to be—not as big, and too much silt."

Generally the herder has a good food selection. He keeps a pencil and paper handy to jot down items he hankers for. He feels eggs, ham and freshly butchered mutton are safe without refrigeration. Canned fruits and vegetables, dried beans, potatoes and cold cereal are on his list, flour and yeast are too if he makes his own bread. Some men are talented with sourdough—hot cakes, biscuits and bread.

Onetime herder Bob Hughes, Firth, Idaho, said herders take special precautions to protect their sourdough on cold nights.

"If they thought it'd work to sleep with the sourdough jug, they would," he said. "But a man will wrap it in his sheepskin coat, put a blanket over that and then put it in his box. On thirty-below nights after the fire goes out, everything in the camp freezes. In the morning canned goods will be popped. But if the sourdough jug is okay, a man feels all right."

Sourdough cooks claim that quality sourdough products come from strong, old starter, "the rottener the better." And frequent use improves sourdough flavor. Virg Shinn, seventy-five, Taylor, Idaho, who has herded since he was fourteen years old, said he sometimes would use his sourdough up to three times a day.

"In the morning, I'd make hot cakes. At lunch, I'd make my bread, and at night, if someone came by, I'd make biscuits. In hot weather if you're using it that often, a batch o' bread will raise in an hour. Raise and run all over."

Shinn said herders know not to wash sourdough jugs.

"If you use a sourdough jug for a year, or year and a half, without washing it, you'll really have good dough."

In spring, tiny wild strawberries make delicious snacks for herders to pick in their ample free time. In late summer, wild raspberries color meadows in startling red.

In fall, chokecherries, serviceberries and huckleberries are available. After a herder eats his fill he can dry the rest in the sun for raisins, boil a batch with water and sugar for hot-cake syrup, or make wine.

In the old days, herders recall there'd still be snow in midsummer on high places like southeastern Idaho's Flagg Knoll. Herders would put canned milk and snow in a lard pail, add rock salt and, after endless turning, have ice cream.

Piñon pines are uncommon in Idaho, but two clusters of them sit near sheep trails in the southeastern area. Herders gather cones, put them on canvas, then stomp out the flavorful nuts. Old-timers say yields are not consistent—alternate years seem to be most abundant.

Whatever he has on hand, the herder usually is eager to share it. People are a welcome treat for him and he treats a visitor as carefully as good china. The coffeepot is ready for instant warming. And since the stove is fired up, how about lunch?

Hospitality and helpfulness are occupational traits. Where isolation is a day-to-day reality, a sense of interdependence develops. Whether the herder is pulling a stranger's pickup out of mud or checking on an ailing friend, he hopes the goodwill will return to him if ever he needs it.

At Nick and Emma Padilla's camp, herders from a wide area gather for conversation and Emma's cooking.

Emma, fifty-nine, has herded with her husband for thirty

years in Colorado and Idaho, and often has cooked for lambing crews. She grew up in orphanages in Colorado where treatment was harsh. But cooking was an area where she could win approval, even awards.

A mile from Emma's camp a wonderful aroma seeps through the pines. It's bread day and Emma is taking a golden round loaf from the oven. Pan rolls cool on the counter. On the stove's top, beans and spanish rice simmer.

When friends from town bring fruit she cans it in her small pressure cooker. When Nick shoots a deer she makes jerky. First she soaks the meat in salt water, then hangs long strips in the pines for the sun to dry. She has a favorite trick to liven up pancakes and drop biscuits—she substitutes beer for water in the biscuit mix.

When a visitor comes by, Emma stops her work, wipes her hands on a dish towel made from a flour sack and insists he come in. She pours coffee, then urges the guest to have helping after helping of fresh warm bread. It is light as angel food from two risings and a half hour's kneading.

While she talks she mashes up hot peppers with a stone metate to make hot sauce for the beans.

"Mash a can of tomatoes, dice a small half onion, add the peppers, a palmful, dice up garlic and put in a little grease, add water and simmer," she explains.

"Have you seen the scissors, Emma?" Nick calls from outside. A handsome young herder from a few miles away is seated on a log with a towel around his neck. Nick is going to trim his hair. "Oh well"—Nick grins—"can't find 'em. I'll take it from the roots."

"Tell him to stay for lunch," Emma calls.

In July herders begin to move bands to the high country. Terrain is too rough for pickups to move camps so herders pack out with teepees. Then the camp tender delivers groceries by pack horse.

In the teepee the herder's life seems to be even humbler. Yet he has packed out an adequate supply of clothes and provisions. Pack boxes (wooden cupboards) ride in the panniers (pack sacks) with an assortment of spices, utensils and packaged foods inside. The herder sets the cupboards inside his teepee next to the bedroll. Cheese, cookies, macaroni, eggs and canned goods continue to be part of the diet.

Some herders have a wooden grub box, a personal collection of items which make for self-sufficiency—instant fruit drinks, coffee, cooking grease, cups, a small frypan, canned meats, a hunting knife, fishing lines and hooks and matches.

In the mountains the herder can easily find pine pitch, which he shaves to use to start fires. Even used with green wood, the pine pitch bursts into flames like kerosene. Trees also supply the herders with pine gum, which resembles regular gum without the sweet flavoring.

As soon as he pitches camp, the herder gets to work improvising furniture for the pack camp. A fat log, sawed to two feet high, makes a good breakfast chair. Slender logs nailed to fatter cross pieces make benches and tables or even an elevated bed between two trees. A log corral surrounding the teepee and cooking area will keep out the horses.

Fresh game and game birds may start to appear in the dutch oven. Grouse are so slow and unsuspecting that a herder can knock one out of the trees with a walking cane if he doesn't have a gun along.

Herders dread a bear in the sheep, but if a herder shoots one he'll have fried bear for many days. Many herders like this rich meat and think it resembles pork.

Trout line up on the herder's breakfast plate. One enterprising man transferred live fish from the river into a small pond in front of his teepee. When he wanted fresh fish he tossed in his line right from his front door.

In some isolated ponds, fish are so thick that herder and camp tender can stage an aquatic roundup. This method works, herders claim, when a man doesn't have a fishing pole and the tantalizing sight of rainbow, cutthroat or golden trout milling in schools is too tempting to ride past.

One man, afoot or on horseback, wades into the water upstream and scares the fish. When they try to escape, the downstream man either clubs them or snares them with a makeshift net.

One herder tells an extraordinary story about a Basque whose dogs understood the game and took the upstream post. They would wade into the creek and start to swim downstream where their master waited with an onion sack. This scheme would net the man seven or eight fish in a few minutes.

Resourcefulness is particularly strong among Basques and helped earn them their excellent reputation. Whether in teepee or camp, the Basque is well prepared. An ample wood supply is chopped and stacked outside. In hottest midsummer, with temperature in the nineties, the Basque still replaces wood he's used for cooking, and his tall stack testifies he is ready for any sudden weather change.

Every vessel in his camp holds water—teakettles, pans, empty wine bottles and jars. He can aggravate the camp tender by wanting his formidable wood and water supply moved along with the camp.

Coffeeola is a common Basque breakfast and the herder will get everything ready the night before so its preparation doesn't slow him down in the morning. He sets out wood for his fire, fills his water pot and measures his coffee.

In the morning when the water boils he makes coffee, then pours in three beaten eggs. To that he adds a little cream and sugar. Some Basques break bread or cheese into the cup to complete the liquid breakfast.

Basque bread is renowned in the West, though fewer herders trouble with it today. Buried on coals a couple of feet down in a dutch oven, the bread cooks slowly for several hours while the herder tends his sheep. When he returns, the herder hauls up the dutch oven and peeks at his bread. He touches it to his nose, the story goes, and if he burns his nose he knows the bread's inside is still doughy. He reburies it for an hour and when he tries again it is just right.

One rancher recalls a Basque herder who always waited for his visits to slice a fresh loaf. Then with some ceremony he would give his boss the first piece and announce that those who got the first slice got boy babies.

Some herders recall watching Basque herders slash the sign of the cross on the loaf's top, and then hand the first piece to the valuable dog.

Guests who sample Basque dutch-oven meals rave about them. A second dutch oven, containing mutton or lamb, onion, spices and vegetables, can be buried in the coals along with the bread and the food is ready for eating when the herder returns from his chores.

The few months when herders aren't with the sheep they find lodging in town—rooming houses, trailers or hotels. Ranchers keep some men on year-around to feed hay. Some herders go to relatives' homes, some find farm or other work during the lag and some have a regular summer-winter schedule which involves two jobs. But all too often the gap in work is an expensive time. Bars provide a ready opportunity for socializing in somewhat unfamiliar towns, and money drains off fast.

The simple sheep camp looks good to the returning herder. Headlines the radio blares about stock-market dips, power blackouts and airport terrorists have little significance for him there. While he minds his sheep he is assured a good firm bed, plenty to eat and peace and quiet.

"When I'm in town I'm down and out," one herder said. "But when I'm up here"—he gestured across a green hill that overlooked a river valley—"when I'm up here I feel good."

IV

He Maketh Me to Lie Down in Green Pastures

When the valley winds carry aspen-tree wedding bouquets and the bitterroot plant turns emerald against the melting snow, then sheep begin to move.

Ranchers try to coincide lambing with spring's coming. They hope that snows will recede and rich grasses will appear for their lactating ewes and the young lambs.

Late spring means economic hardship—it takes more feed to maintain heavy-milking ewes. So when the foothills turn green the herder gathers his band and moves them out of the pens.

Baby lambs are hard to herd. They remember the last place they suckled and try to return to it even when their mothers are moving forward. If one gets past the herder he'll give a horse a good race back to the pens. As the fuzzy body dashes by, others may decide to follow.

If several men are helping, men afoot try to head off lambs by tossing chaps or coats in their paths. Dogs are watchful for escapees too.

Noise from the sheep jars the drowsy hillsides awake. Confused lambs wail and worried mothers blat as they get separated in the milling and shoving. An anxious ewe will touch her nose to several lambs, calling loudly, until she finds her own.

Emlen Hayes, Howe, Idaho, rancher, observes that sheep are color blind.

"If a mother with a black lamb gets separated from her lamb she will have as much trouble finding it as mothers with white lambs. There he'll be, standing out like a sore thumb, and she'll walk right by him crying for her lamb. The herd doesn't discriminate against the black sheep either."

Ranchers group together in one band yearling wethers (castrated males), which weren't sold the previous year because of size or market conditions, and yearling ewes, carried over to become breeding stock. If a rancher lambs early, pushes the lambs with rich feed and sends them to an early market, he may have a band of dry ewes too. So the herder has in his charge either yearlings, dry ewes or ewes and lambs. Herders find the latter easiest to herd. Ewes with lambs are happier and settle down better. "It's against nature," a herder said about his band of dry ewes. "These gals know they should have lambs with them and they don't settle down good."

Many ranchers use black sheep as markers, one or two per hundred whites, so the herder can make quick random checks. Especially where predators are a problem or where sheep are trailing in dense country, the herder frequently counts his blacks to get an idea if he is losing sheep.

Some bands have a lead sheep or goat tamed to walk on a leash. When loading sheep into trucks or moving them into chutes, the herder walks the lead sheep past the others. They follow.

Like all young things, lambs find great joy in moving. They run, leap, whirl in midair and chase their stubby tails. In groups of two or three they play tag and follow the leader. They have a gracious way of dropping to their knees to suckle. Few men could be untouched by their clownish behavior and trusting faces.

"A lot of herders take as good care o' lambs as they would a bunch o' kids," said Rube Moses, Blackfoot, Idaho, who herded forty-five years. "You never heard a man cuss like when a herder finds a dead lamb that the coyotes got."

A case of exaggerated herder loyalty happened in Montana. A Basque tending ewes and lambs tried to water his band, but a Mexican tending yearlings blocked him from the creek. Aware that a mix-up with the yearlings could result in lamb losses and injuries, the Basque didn't drive his band into the water while the Mexican stayed. Next day, the Mexican blocked him again. Water was more crucial to the lamb-ewe band than to the yearlings, and the Basque got madder. The third day the scene repeated.

When the Mexican mounted his horse to ride off, the Basque pulled his rifle from his scabbard and shot him in the back. The Basque served time in a Montana prison. Later he was paroled to a big sheep outfit.

Did the man have a bad temper? No, say people who knew him. But the old-time good herders loved the sheep so much they didn't let anything interfere with their welfare.

A rancher recalls the dedicated herder who embarrassed him with a neighboring sheep outfit. Once a neighbor sheepman was visiting, helping to load lambs. The neighbor allowed his dogs to run after lambs, which made the conscientious herder madder and madder. Several times the herder warned the man to watch his dogs, but the neighbor paid no attention. Finally the dogs chased one lamb so far away that even a careful search couldn't turn it up.

Next time the rancher and his boss passed a band of the neighbor's sheep the herder ordered the boss to stop. He walked through the herd, selected a fine, fat lamb and carried it off to his boss's pickup. He didn't say a word but his determined look read "a lamb for a lamb."

Sometimes when trailing, a herder will pick up a weak straggler and carry it himself. Unlike sentimental pictures of Jesus, the herder does not cradle the lamb in his arms, but hauls it up and totes it by a hind leg.

Lambs see and walk almost from birth. Their teeth break through the gums within a few days and lambs are able to munch grass and hay when only a week old.

People who raise bum (orphan) lambs dispute the common notion that sheep are stupid.

"I think it's like any other animal, the more it's worked the smarter it gets," said a woman who raised a weak, February twin in her kitchen. Her lamb went along on car rides, snugged up to her chin and was almost housebroken when weather warmed enough to put him in the backyard. Each morning when the woman left home she enclosed the lamb in a kitchen area and covered one corner with newspaper. The lamb confined his messes to the newspapers and never had an accident in the car or on laps. When he needed to be put on papers he gave his substitute mother a desperate, meaningful look and wiggled his hind end.

A bum, who gets a warmed bottle of calf-replacement formula every four hours for the first couple of months, quickly learns his name and wags his stumpy tail when he hears it. He comes running when called. He can be an ideal pet for very young children because he is gentler than a rambunctious puppy and stands quietly for hugs.

Another woman who raised a bum tells how her lamb joined her and her children for daily walks around her country neighborhood, following at her heels without a leash. That same lamb invented a game of butt the chair on her children's gym set. He would aim, run at the chair swing, then watch it move to and fro. That lamb also became skillful at a game of roll the ball with your head.

One bum took up chasing cars with the family dogs. His teetotaling family kidded that their lamb may have inspired some drunken driver to discard his habit for good.

Early months are important ones for lambs. Their fastest growth spurt occurs in the first thirty days. They are efficient feed converters when very young, and can easily gain eighty pounds in the first eighty days. Ranchers push for good gains in the first four months because of the favorable feed-conversion ratio. After that gains are slower, and lambs stunted in early life may take a long time to achieve market weight.

The first month is also a critical time for disease. Parasites, viruses and pneumonia can claim young lambs en masse and wither profits.

Pneumonia moves in with stress or bad weather. Herders can monitor stress due to trucking or trailing but no one can predict when cold late storms will bear down on mountains and deserts and trap flocks belly deep in snow. Young lambs barely in wool are vulnerable to prolonged winds and moisture, and a bright outlook for a healthy lamb crop can dim during a bad-weather spell.

Worms bring down a lamb's over-all condition and tend to be a problem in enclosed areas. Once on the range, the problem lessens.

Most outfits administer Aureomycin to prevent scours, a dysentery which exacts a high death toll. Yet, a few varieties resist the drug. Coccidiosis, an intestinal parasite spread through the droppings, drags down sheep. It is more common than before, ranchers say, and they theorize that certain germs may be building resistance to drugs.

Navel disease strikes when the navel is still wet and throbbing, but doesn't show up for a couple of weeks. Then it will viciously wipe out a lamb crop. That disease requires the now standard practice of swabbing newborns' bellies with iodine.

When sheep are no longer confined, the sickness threat eases. And vitamin-rich grass popping up on hills and wide stretches of early flatland pasture has a salutary effect on them.

Sheep like young grass. When stalks get longer, sheep pass up grass for other food.

A good portion of their diet is browse, or woody plants. They love bitterbush, called buckbrush by stockmen. Sheep love its yellow flowers, which have an orange-blossom fragrance, and they also like the plant's foliage.

Sheep eat sagebrush and its smaller ground-hugging cousin, mountain sagebrush. They like serviceberry, snowberry and chokecherry bushes.

In spring, meadows turn purple with crow's-foot, a flowering plant. Sheep eat the lavender blossoms and quite a bit of the foliage. Dock, a plant with wide, deep green leaves, paints the hillsides yellow when it blooms. Sheep like the plant until it is two or three inches high, then they pass up mature leaves and wait for its yellow sunflower to appear. Sheep feast so greedily on the flowers that their wool starts to take on a yellowish cast.

Sheep "fill up like ticks," herders know, on cow cabbage, a tall broad-leafed plant with full, white flowers. The plant is so rich a whole band will fill up on a quarter acre of it, then lie down and rest contentedly for hours.

Lambs continue to suckle, along with their other feed, up to shipping time, about five months old. However, a lamb could lose his mother as early as three weeks and still survive.

Though the land is abundant with friendly plants, the herder must be watchful that sheep stay away from toxic ones.

A field of wild beans, lupine, can be deadly. Assorted varieties grow on desert, foothills and at higher elevations. They grow in colorful masses of blue, blue-lavender and white, and plants have small pods. They are a well-known spring flower throughout the West. Sheep unaccustomed to the plant are particularly susceptible.

One rancher said he once lost four hundred sheep during a herder's lunch break. The rancher had warned the herder not to let the sheep rest in a spot near a spring where the lupine grew thick. But the herder didn't heed the warning, and by the time the rancher came along sheep were dying. Some other reported losses from lupine have been as high as a thousand head.

Death camas, a slender plant with cream-colored flowers, comes out in spring before other range plants, so herders watch for this toxic weed in early season.

When leaves are unfolding on chokecherry bushes, that plant is highly dangerous. Again in fall, when frost causes the leaves to fall, they are harmful. If sheep get too hungry they will munch leaves off the ground.

Chokecherry leaves cause bloat. Sheep bloat too if they overeat on leafy alfalfa. Ranchers say there is no satisfactory treatment for the condition, which takes sheep fast. The movie device of sticking sheep in the ribs with a knife to relieve pressure rarely works, ranchers says.

Halogeton, a plant that looks like Russian thistle, has wiped out entire herds. It moves in where ground has been disturbed and now thrives in many desert and mountain pastures. Herders can feed dicalcium phosphate when they know that sheep will feed where halogeton grows, but after the fact treatment does little good. The weed causes bloat and sheep die quickly.

An attractive menu awaits the sheep, but the herder must go before and search the meadows. If hazards lurk in the appealing fields, then he has to steer his sheep another way.

V

He Leadeth Me Beside the Still Waters

A second-generation sheepman remembers a story his father told about sheep and their preference for still water.

The old man ran sheep on the Utah desert. He moved them great distances, from Idaho to Utah, then through Utah almost to the Nevada line. They trailed over powder-dry land where the desert sun boiled moisture out of vegetation. Once animals went without a drink for three days.

In the midst of this parched stretch flowed a swift, dependably full stream. From a distance the sheep smelled it and began to run. When they reached it they ran up and down beside it but would not drink.

The stream ran on to some quiet but deadly alkaline ponds and the sheepman had to send a lead man and dogs ahead to hold back the sheep from the dangerous water. It took hard work, vigilance and good dogs to discourage the craving band. Eventually the animals gave up trying to get to the ponds and gingerly slaked their thirst in the high, swift stream.

Sheep's dislike for fast water can be a problem during spring trailing when streams swell with mountain snow melt. Sometimes a trail crosses a tugging, roaring creek. Even though the

herder selects the quietest place to cross, sheep can panic. They line the bank, mill, sniff and stomp their front legs. Nothing will budge them.

Sometimes a mounted herder can cross the creek, then give the "brrrrrrrr" salt call to trick his herd to follow. It works if the herd's salt hunger exceeds its fear.

If the herder can persuade a lead sheep to cross, others will follow.

Where water is very high and fast, men sometimes have to construct a makeshift bridge from available logs. Then one problem still remains—the herder must coax the sheep onto the bridge.

Dogs can be useful in moving sheep across water, yet a frustrated dog can inflict harm. In tight situations an overwrought dog can be too rough on young lambs and draw blood and break legs. A broken leg heals if a herd is stationary, but when sheep are trailing the herder must splint the leg or have the lamb hauled. If the herder applies a splint, chance of infection increases.

Sometimes patience is the only way to deal with scared sheep. After a band gets fully used to the water's roar, a few may decide it is all right for watering and wading into.

Many outfits trail out to rich spring pastures in April, then trail back in May for shearing. Often herders choose a lush path along a creek where aspens and cottonwoods provide good shade. Overheating affects quality of fleece—when animals get too hot, grease in the wool surfaces and collects dirt.

Sheep are somewhat thirstier before shearing, when heavy wool causes them to sweat more. They sweat where chest and forelegs attach, in approximately the same place as a person's armpits.

A nursing ewe requires two gallons of water per day—other sheep need about half that amount. Stockmen try to avoid hauling water to sheep before shearing because animals get dustier around troughs than they would around stream banks. If an outfit sells on a "clean basis" the buyer extracts a sample from every wool bag, tests it for cleanliness and pays accordingly. If sheep look white across the back when they enter the corrals it usually indicates they will shear clean and wool will bring a good price. If unfavorable conditions precede shearing, sheep will have a slate cast to their fleece.

When bringing sheep in for shearing, herders watch the skies apprehensively. Sheep should be dry when sheared. And after shearing, naked sheep are vulnerable to temperature drops.

Some sheepmen say the worst thing is a cold wind that whips up after shearing on a warm day. Ewes chill quickly and can get colds and pneumonia. Even forty-eight hours of good weather after shearing will help them acclimate.

Shearers arrive at the shearing corrals in a bus, followed by a two-ton truck that houses the shearing plant. The crew consists of about eighteen shearers, two fleece tiers and one wool tromper. Most crews today are Mexican-Americans from the Southwest, with a few Mexican aliens along too.

The men set up quickly; tents and sleeping cots for the crew go up and portable canvas or plywood floors for shearing go down. Above the floors the men erect protective awnings or a tent.

A generator mounted on the truck provides electricity for the operation. A half-horsepower motor above the shearer's head connects to a fifteen-foot metal shaft which takes power to the handpiece.

One man is busy full time at a power disc keeping blades for the handpiece gleaming sharp. Sheepmen usually count on losing a few sheep from cuts, internal injuries due to rough treatment, stress or unfortunate weather.

When tent and machinery are in place a man brings the company's own lead sheep on a leash through the corral. The client's sheep follow into the shearing area.

Shearers work in a line under the awnings. When a shearer chooses a ewe he grabs her, pulls her to his area, puts her on her back and binds her legs. He runs the blade over her in a circular manner and tries to leave the fleece in one piece.

The fleece tier collects the shorn coat and ties it with paper twine, which later will wash with the wool. He hands the shearer a token, usually a washer or wooden slug. At day's end the shearer's wages will be based on how many slugs he collected.

A fast shearer can do a sheep in two and a half minutes from the time he has her in tow. Shearers average 100 to 150 a day, but herders tell of men who can do 200 and 210.

The fleece tier delivers the wool to the tromper, who puts it in a large vat to compress it for sacking. In the old days trompers suspended the sack on a frame and folded its edges over an iron hoop to secure it. Then the tromper got in the sack to stamp down the fleeces, an aggravating job because ticks fleeing the wool jumped onto the man. Though not harmful to man, sheep ticks are crawly, pesky bugs.

Before sophisticated electric-shearing crews took over, men using hand clippers did the work. A good hand shearer could do 100 to 150 animals per day. A few hand-shearing outfits still operate. Outfits that lamb on the range or graze colder regions prefer hand crews because hand shearers leave more fleece.

The shearing company charges the sheepmen about $1.10 per ewe, $2.20 per ram. Ewes weigh about 150 pounds and rams about 300 pounds, but rams do not produce twice as much wool on their bigger frames.

In Idaho sheep average ten pounds of wool per animal before shrinkage. Animals of better breeding produce more. Other variables are feed, location and breed. Sheep exposed to severe weather do not grow more wool for warmth—abundant fleece goes with good conditions and mild winters.

Rambouillets, the breed to which all white-face sheep trace their ancestry, produce the finest wool. Wool is graded against the Rambouillet standard. Some breeds produce a coarse fleece that makes carpet wool. At one time those fleeces sold low, but now sell high because of their scarcity.

Buyers have a strong prejudice against black wool, and the more black fibers found in wool the less it brings on the market.

The sheep emerges from shearing a different animal. She went in a round and pretty storybook animal. She comes out a bare, angular cross between goat and small cow.

As she passes through a narrow chute to leave, a herder stamps his employer's brand on her. He may give her a quick spraying with insecticide.

Sometimes the sheep stay overnight in the corral after shearing, huddled for warmth. But usually they need to move out to make room for other bands.

The herder isn't bound to follow the creek back. Spring grass is high in moisture, and if days are cool sheep can go a week without water.

But probably he will. In spring, mother ducks hide their

ducklings behind willow branches that dip into the water. Tall sandhill cranes visit the creek at sunrise and twilight, sounding their odd gargling call. Does and fawns water within the herder's view. Snowberry bushes in pink blossom decorate the banks and catnip plants offer dull, powdery leaves for making tea. The herder likes the familiar activity along the creek. The sheep likes its gentle water.

The herder gets well acquainted with animals. He depends on and appreciates his dogs and horse in a way mechanized society has forgotten.

His working partnership with the dogs little resembles the common owner-house-pet situation; the dogs are at least equal members of the sheep-care team.

"You can't herd without a dog," Ab Waters said. "Oh, maybe the first day the sheep won't know the dog is gone, but as soon as they find out, it's like a bunch o' kids when the teacher's gone. They get real mean and full o' mischief and they won't mind."

Every herder can tell of a marvelously responsive dog who could almost anticpate his master's wishes. Such a dog could be sent up a hillside alone with the sheep. When it got a distance away, the dog would turn and look back at the herder for further instructions. The herder would wave his arm the direction he wanted the sheep turned; the dog would race off to obey.

Retired herder John Hunt had a dog who would hold the sheep back from the hay until his master gave permission to let them eat.

"In the days when we fed from a hay wagon I could scatter the hay over a big area and the dog would keep the sheep back until I was all done," Hunt said.

Hunt could leave another dog, Jerry, in charge during mealtimes.

"We'd be a-bringin' sheep out of the mountain, trailin' on the road. When the camp tender said it was time to eat I'd tell Jerry, 'You tend 'em now.' He'd hold 'em right together on the road until I finished dinner, then I'd let 'im know it was all right to let 'em start movin' again."

Accounts of animal heroism come out of the camps. One

dog ran twenty-two miles home to get his master's wife after the man was bitten by a rattler. The herder had trained the dog, when it was a pup, to return home on command. The rattler struck in the morning and by noon help reached the man.

While still pups, dogs begin to help their mothers. Their natural abilities, combined with the example of older dogs and discipline from herders, make a working dog.

Border collies and Australian shepherds and mixes from the two breeds are the most common in Idaho camps. But some herders claim that any dog, given the opportunity, can learn to work sheep. At the Padilla camp, a friendly black Labrador retriever helps herd. Ab Waters claims that the best sheepdog he ever had was a cocker spaniel. Some outfits have had success with the big, shaggy English sheepdogs, and a few people use the large white Great Pyrenees.

Dogs love to work. Jack, who works with Basque herder Juan, drops to the ground and whines a protest when Juan calls him off the sheep. When loading sheep into chutes, some capable dogs will leap over the backs of milling sheep, then jump onto the stalled front sheep and bark or nip it into action.

Rube Moses, who herded and tended camp for forty-five years, recalls his dog Rusty. Rusty was with him sixteen years, and though his outfit trailed sheep over roads for many seasons, Rusty never let a lamb get hit by a car or truck. And Rusty would fetch horses as easily as he worked sheep. He could lead a horse, holding the halter rope in his teeth.

"In the spring when the thistles got bad, Rusty's feet was gettin' tore up. I cut up an old sheepskin coat and made some moccasins and it helped him out," Moses recalls.

In the days before packaged dog food, Moses said, the herder cooked up pancakes and an egg and some bacon for the dog along with his own breakfast.

Dogs can be very protective toward sheep. They bark to alert the herder about bears and coyotes and try to run off predators. It's unusual, but coyotes in a group have turned on a pursuing dog and killed it.

Emma Padilla raised a large Great Pyrenees that fussed over sheep. One year two bum lambs Emma raised were grazing a field near her camp. A coyote on a nearby hill moaned. The dog ran off, rounded up the lambs and pushed them under the camp. The lambs preferred the nice meadow and wanted to escape, but the dog pushed them back with his nose, and kept them there all night.

A friendship blossomed between another bum, Whitey, and the Padilla's dog, Bob. On moonlit nights their silhouettes showed them snuggled together, asleep. If it started to rain the pair got up and ran under the sheep camp and lay down together once more.

But there was rivalry between them for Emma's affection. Once when Emma was helping load lambs into a chute, Whitey noticed Bob sitting between Emma's legs. The jealous lamb ran up and slammed the dog with its head and sent the dog flying into the willows below.

Whitey also competed with Bob for dog food. Nick hid dog food in the pickup, but Whitey liked to jump into the pickup for rides and often found and diminished the supply. When Nick wanted to feed Bob he walked a distance away while Emma restrained Whitey. After Nick had filled and hidden Bob's dish he would yell "all clear" and Emma would release Whitey. Often Whitey found Bob and the dish anyway and helped Bob clean up his dinner.

In the herder's free time he often visits with his dog. During such unstructured hours Emma taught one dog, Rusty, to sing. Their favorite duet was "How Much Is That Doggie in the Window?" Rusty contributed his "woof, woof" in appropriate measures. He enjoyed the game enough that in time all Emma had to do was raise her finger a certain way and Rusty knew to start howling. Around the cookhouse herders dropped by to request "Make Rusty sing." Emma would move to the doorway, lift her finger and Rusty would break into tuneful moan.

Though the horse's role isn't as dramatically vital to herding as the dog's, the herder relies on his horse to help him carry out day-to-day chores. He treasures one that is willing, hardy and dependable. Herders often are in true survival conditions where a horse's good judgment and surefootedness can stave off disaster.

Rancher Allan Thompson tells of coming through Bear Creek Canyon at night on a horse named Julie. In places the trail is only inches wide and footing is slippery and rocky. A misstep would mean a fall hundreds of feet into Bear Creek.

Thompson had taken a wrong turn on a trail and found himself on a high ridge in darkness. He wasn't equipped to camp overnight. The horse had proven herself many times before.

"I never get scared on a horse, but that night, coming across that ledge, I was shivering. If I hadn't had such trust in the horse I couldn't have done it," he said.

One outfit undertook to remove horses from its operation and replace them with motorcycles. A young man took over his father's business, and having done some hill climbing with bikes, thought they would be a trouble-free alternative to horses.

Herders scoffed at the young man and told him he underrated the terrain. He couldn't persuade any herders, even young ones, to take a bike instead of a horse. But he thought if he demonstrated one successfully for a season, herders might come around the next year.

The sheep left in May. In June the new boss toppled over a mountain on his bike and ended up in the hospital with multiple injuries. No one mentioned the plan anymore.

Horses in the camps become quite domesticated, sometimes pesty. Anne Thatcher of the Mays sheep-ranching family tells of a horse-herder team who shared a love for pancakes. Sometimes the horse would poke his nose in the sheep camp and snitch them off the griddle when the herder wasn't looking. Thompson remembers a horse who thieved fried potatoes. When it smelled spuds cooking it came around, and the herder's supper disappeared from the pan.

James Mays had a knock-kneed, pigeon-toed palomino named Captain who was mechanically clever and could untie knots, open gates and work water hydrants.

Mays' camp jack hid oats under his pickup at night so horses couldn't get them, but Cap maneuvered them out anyway. The camp jack decided to foil Cap; he put the oats on the pickup's seat. During the night Cap figured out how to pull the handle. He opened the door, dragged out the sack and fed himself and his friends.

In an area where his outfit had to truck out water, Cap

watched the camp jack pull the hose off the truck and put it in the tub for horses. One day, the camp tender said, he guessed he wasn't fast enough with the water because Cap took it upon himself to reach up and bring down the hose. Cap couldn't replace the hose, and water went all over.

Around the ranch, Cap used his teeth to turn on all the water hydrants. "But he never shut anything off," Mays complained.

Cap liked to stick his rather large head into conversations. It made herders laugh when Cap's entire head poked in the sheep-camp door.

Though twenty-four years old in the summer of 1977, Cap still insisted he be lead horse when his outfit packed out to the high country. On the return trip the herder tied him while he checked some sheep, and when he returned minutes later Cap was on the ground, dead from an apparent stroke.

"I felt good about it," Mays said. "He started out in the hills and that's where it ended."

Cats can be good pets and useful too. They can keep down the number of mice which get into groceries and grain.

Ab Waters adopted a stray which never became secure about its portable home. On moving day the cat would flee into the camp and hide out of reach, afraid it was going to be left behind.

Emma's cat is Negrita. Emma demonstrated how Negrita likes to dance. She held the animal's front paws in her hands.

"Dickera, dickera, dickera, dee," she sang, "dickera, dickera, dickera dee. See? She likes it." The cat, face impassive, looked off into the distance.

A Basque legend says that Euskera was the tongue spoken in the Garden of Eden, and all animals understand it. Some who have observed Basques at work think their communication with animals is special. Sheep people aren't unanimous in acclaiming Basques as excellent herders, but almost everyone agrees that when a Basque does distinguish himself his talent with animals has a lot to do with it.

Rube Moses remembers a Basque named Virg. Virg's dog understood his master "just like a person understands his friend."

"I've seen a lot of good dogs but I never saw one like that. Virg would hold the dog in his arms and talk to him and point with his fingers. Then he'd set the dog down and that dog would run up and turn the band of sheep right where Virg wanted 'em.

"He never raised his voice to his dogs and he never laid a hand on them. And when you went to cut out sheep, you never had to wait because his dogs were working every minute."

Virg's horse, a homely, black jug-headed horse, became almost a house pet.

"Virg and Jug, them two lived together," Moses said. "One day when Virg was packed out in a teepee, I came to help him move camp. We looked all over for Jug, but couldn't find him. The mosquitoes were thick, and we were gettin' aggravated.

"Finally I noticed the dogs standin' down by the teepee. I says to Virg, 'Ya don't spose ole Jug has gone and got in that teepee?' 'Naw,' says Virg, 'he can't get in there.'

"But I went down to look and when I opened the flap there was Jug, layin' out sound asleep, his head on Virg's rolled-up sleeping bag. Virg kept his teepee sprayed good and I guess Jug decided it'd be a good place to get in out of the skeeters."

The herder feels friendly with wildlife too. It's always a thrill, a herder said, to come on to a clouded pool, stirred up by the recent exit of elk, bear or moose. The herder becomes expert at track identification and often indulges in leisurely walks to follow deer, cougar, elk or bear prints. Many men can name at a glance the number and sex of animals recently on the path.

One man said he cusses as much when he finds a dead elk or deer calf that coyotes have killed as when he finds a dead lamb.

The small animals can entertain. Rock chucks, badgers and rabbits scurry along the dirt path. The imprint of green willow branches tell where beavers have dragged their building materials.

One herder looks forward to the high serviceberry bush near Castle Rock where he always sees deer. Every year, for as long as he can remember, whenever he rides past that bush he scares up deer. A cool breeze comes off a nearby ridge and the herder thinks does must teach their fawns about the place. Always at least one deer is shaded up there.

Thompson recalls two episodes when he tried to get better acquainted with elk. Once a herder mentioned to him that in all his years in the hills he'd not once seen an elk. A short time later as Thompson rode through a meadow he came upon three bull elk. He decided to round them up and herd them back to camp for his friend to see. His horse reacted skittishly to the idea, but Thompson coaxed her into top speed. He tried to bend the elk toward camp but "it was like running next to a freight train. It really impressed me how powerful they are."

Another time he found a pretty elk calf lying in the brush. He rode right up to it and it peered up at him with round brown eyes. He wondered if it might like to play, and he decided to catch it.

The best way, he figured, was to fall right onto it. When he landed, the calf was bigger than it seemed. It lurched, fought, flailed, tore his shirt and pants and ran away.

Whether watching wild creatures or training sheepdogs the herder's job gives him time and opportunity to learn about animals.

VI

He Restoreth My Soul

The weekend was a disaster. I spent the Memorial Day holiday in the mountains on a picture-taking expedition. It was a shakedown cruise, too, for my longer stay in the sheep camps later this summer.

It started out promisingly. I took it for a good omen that Lark, my three-year-old mare who only recently was broke to ride, loaded into the horse trailer without difficulty. Despite the skittishness she feels toward people (she was a range colt and never was handled until I bought her at age two), she shows some trail sense and I looked forward to trying her in the hills.

In my customary manner I planned an early departure and got instead a late start. The sun was threateningly low by the time I got into the mountains, leaving little daylight to search for a destination I was unclear about. I planned to camp along a creek beneath a hill where Thompson's Basque herder, Juan, was tending his band, then go up the hill early next morning to get some dawn pictures.

On an incline on a gravel hill, the truck quit. This was disappointing but hardly surprising—machines and I hate each other, and whenever I think I'm successfully dominating one it boils over its engine to put me in my place.

It was an unfortunate spot to be stalled. My foot held down the brake so I wouldn't slide backward, but I couldn't stay like that indefinitely. Going forward didn't work either despite several attempts to restart the engine. Perhaps I could back down the hill gradually and just spend the night in a tree-lined spot I saw at the ridge's base.

The leeway for error was small and I was still trying to work out mentally the rule for backing trailers when I saw one trailer wheel slip over the hill's edge. I imagined Lark, balanced precariously, deciding never to load in a trailer again.

My right thigh and calf felt weary from pressing on the brake; the quiet woodsy spot now looked ominous. I worked out an emergency plan—if the brake failed I'd jump out as the whole outfit went over the ridge. The hour was a little late and the road a bit deserted for someone to come to my rescue.

But, someone did. For being rescued, Idaho is the most dependable state I've lived in. Its residents have a keen instinct for finding stranded motorists, and they are unfailingly willing to help.

A young man in a cowboy hat hopped out of his pickup, trotted over, reached into my truck and seized my hand.

"My name's Blaine Simmons. Looks like you need some help."

He tied a rope to my bumper and towed the whole rig to a level spot. Lark nickered with relief.

Blaine returned and leaned against the truck.

"Where are you headed so late with that horse?"

I told him the general area where I was headed and that I thought I'd recognize the spot when I got there.

He shook his head. "Hey, have you noticed all these cars up here tonight?"

"Several passed me earlier. Where are they all going?"

"Looks like the biggest beer party of the year. Since they closed the Idaho Falls parks to drinking, the kids come up here. I don't think it's a good night for a woman to be camped alone."

After a moment he said, "Hey, our summer place is vacant. It's about eight miles up this road, after those buttes, if your truck will make it. We haven't yet moved up to start our full-lime farm work. You got a flashlight?"

"No."

"Camping without a flashlight? Well, you probably can use a match to find where you turn the gas on. You'll need it up there."

"No match either."

I made a mental note to be better prepared on my later trip.

"The butane valve is in the back if you want to feel around for it. The door's open, just go in and toss your sleeping bag on the bed," Blaine said and drove off in the opposite direction.

By the time I reached the house the truck's engine was rumbling and hissing. In the darkness I searched for water for both Lark and the truck, but couldn't find any.

But what a break! The house's owners were dry farmers, Blaine had explained, and hadn't yet moved up for the summer. No need to make camp, just go right to bed and get up at dawn to try to find my herder.

My women friends had expressed concern about my camping alone in the hills. I hadn't worried, and besides, I had my gray and white keeshond, Tundra, along. How secure she made me feel as I lifted her, trembling, from the truck.

She cowered by the truck's wheel so I went back for her and carried her to the house. I couldn't coax her into the house's small enclosed porch, so once more I lifted the squirming dog and deposited her inside the door. She whimpered with despair when I walked into the kitchen, but she wouldn't follow.

Enough moonlight shined in the bedroom so I could see where to throw my sleeping bag. I undressed to the sound of mournful calls from the porch.

During the night, I gave up. I couldn't sleep amidst the tortured noises, so I padded barefoot out of the bedroom and over the kitchen's cold linoleum to the porch. I groped in the dark for a mound of fur, picked it up and stumbled back through the dark house. Tundra slept beside my bed, emitting tremulous whines whenever my hand slipped off her head.

A faint morning light next day allowed me to read my watch —quarter to five. As I peered out the window I saw Thompson's familiar brown pickup coming up the hill. If I hurried, I might be able to follow him right to my herder. I yanked on my clothes, dashed out of the house and flung things into my truck. It started!

When I caught Mr. Thompson he was starting up the steep hill to Juan's camp. I borrowed a tin can so I could fill my radiator.

Originally I had planned to hike up from the river with my camera equipment, leave it with the herder, then come back for the horse. I didn't want to risk carrying expensive lenses on a green horse. But after I tied Lark behind the trailer, I decided to hitch a ride up the mountain with Thompson.

The sheep, working their jaws, still lay on the bed ground. The pink morning sky and a ragged ridge were behind them. Juan stepped out of his camp with a hearty wave.

"Cof-FEE?" he yelled.

I'd met Juan on two other occasions, when I'd brought my two-year-old twin boys along, and it was impossible to resist his hospitality. Always he wanted to cook a meal, refill my coffee or share his vino.

Despite his twelve years in the United States, he spoke no English and comprehended only a minimum number of sheep-tending words. I'd enlisted a friend's help in learning enough Spanish to communicate my reason for spending the day with him.

Thompson hesitated a moment as he got into his truck to leave.

"You sure this is all right?" he asked.

"Oh, aren't you? I'll just tag Juan around all day to watch him, and leave at nightfall. When he takes his sheep down to water this morning, I'll get Lark. Do you have second thoughts?"

"Well, I'm about ninety-five per cent sure this man is all right. But he's only worked for me a month."

"We'll be fine," I assured him. My new ability to communicate my business boosted my confidence.

I shot a couple of rolls of film before I returned to the camp for Juan's coffee.

When two people are trying to create a congenial atmosphere but don't speak the same language, the transaction fills up with vigorous nods, "mmmmm's" and "ohh's." A sentence of Juan's incomprehensible Basque or rapid Spanish brought forth happy nods and a "mmmmm?" from me. My carefully enunciated English, spoken in a slightly loud voice, precipitated smiles and "oh, yeah, oh yeah," from Juan.

Intermittently I tried my few Spanish sentences. I explained "*mi caballo*" was "*nervosa,*" in case he wondered why I hadn't ridden up. He looked out the camp door, puzzled. I didn't know the words to tell Lark's location. "Oh yeah, oh yeah," he muttered.

He nodded and smiled when I told him I wanted to take "*fotografias*" of him. But when I told him "*escribo un libro*" he squeezed up his face, baffled. I repeated the sentence. He handed me a magazine. (Another herder, a Mexican, later told me I could have conveyed "I'm writing a book" a better way. While my phrase was technically all right, "*yo estoy escribiendo un libro*" would have been clearer.)

One attempted conversation kept repeating. Juan would motion across the valley, point at me and say, "You. *Muchos terrenos?*"

"No. Not me."

"Yeah, you. *Muchos terrenos.*"

Then he held up four fingers. "Four bahnds lahmbies. You?"

I shrugged.

"Cows. *Muchos cows?* You?"

I kept shaking my head no, and figured that Juan, like some foreigners, assumed all Americans were rich and owned great parcels of land.

Several times Juan mentioned the "*muchachos.*" I told him they were twins, a word he recognized from lambing. Juan smiled greatly, held his hands in front of his chest, wiggled them back and forth and made an mmmmmmmm engine sound in his throat.

"*Hombre*. Peekup. MMMMMM."

I couldn't follow this line of questions and wished we'd find another topic to mutually puzzle over.

When Juan started the sheep down to water I followed behind, clicking pictures and writing down Spanish and Basque words he taught me—sagebrush, rock, tree, bird, dog.

Dog! Where was Tundra? It was the first I'd thought of her. Shame overcame me as I figured out that in my haste I'd forgotten to load her into the truck when I left the summer house. I imagined her, running her heart out to catch up, and now lost in the empty hills.

Juan squatted in the dirt and drew a map with a stick. It looked like Spain. I asked where his home province was and he made a dot for his home, Durango, in Viscaya Province. As I peered over Juan's shoulder I noticed his virile back and swelling biceps and thought of the tales of legendary Basque strength.

"You," he said. "Basque?"

"No."

"*Spani?*"

"No."

"Oh. Basque." He nodded vigorously. Several times since I started work on the book I've had Basques and Basque-Americans tell me I look Basque. Though I always told Juan, no, I was Irish, he insisted that, yes, I must be Basque.

"You. *Muchos terrenos.*"

Here we went again.

"*Hombre.* Peekup." Mmmmmmmm—engine noise. Hands wiggled on an imaginary steering wheel.

It dawned on me! He thought I, and the little boys, belonged to Thompson, along with the sheep and the cows and the *terrenos.*

"No. No," I said. "*Hombre,* peekup, *AMIGO! Mi hombre?* NO! *Amigo! Si!*"

Juan furrowed his brow. Finally the meaning sunk in.

Instantly I saw my mistake. Juan's face changed. His eyes lighted with hope.

"You." He pointed. "Me." He smiled.

"Uh—*escribo un libro.*" I said it loudly though Juan had come closer.

"Lonch? Campo?" He pointed back up the hill. "Vino?" He smiled significantly.

His hands unfolded in front of my face to show all ten fingers. Then he curled and uncurled them four times and pointed to himself. Forty years old. He pointed to me. I repeated his exercise three times.

He held up ten fingers and grinned, apparently delighted at what he saw as an ideal age difference.

When one strong arm swept me up around the waist, I arranged a stern face and pointed indignantly to my wedding band. "Husband!" I said.

"I-daw-ho Fawls?" He smiled bigger.

"No! Uh . . . *campo*. Cows . . . over there!" I pointed. His smile told me it was an unsuccessful lie.

Walking down a mountain, backward, taxes seldom-used leg muscles. In view of my situation I wanted to avoid tripping and sprawling into a meadow and somehow I managed to tread a clear path I couldn't see.

I'd made a bad choice. I might have spent the day with some other herder—a small dissipated man, on the elderly side. A strong young woman would have a chance against such a one. As it was, only a few giant strides separated me from a barrel-chested fortress who possessed a centuries-old tradition of physical prowess.

"*Parlez-vous français?*" I said with hope.

"Mmmmmm? Ooooh." He nodded.

Really I wasn't alarmed, only uncomfortable. A conscientious herder like Juan would make sure his sheep got down to the water before he considered diversions, and beside the creek my truck and horse waited. I fingered the keys in my pocket.

Behind my shoulder the truck's red top came into view. I turned around and began to skip down the hill. Juan, still grinning, missed the significance of the rig parked by the water.

Past the trees, a full view of the truck and trailer made my mouth fall open. Lark was gone. My mind raced, I couldn't imagine—I'd tied her with a foolproof knot. I ran the rest of the way to my outfit.

The halter rope's frayed end under the trailer's bottom corner told the story. Probably she'd gotten her rope caught while trying to eat, and she'd thrashed until the jagged metal sawed it in two. I hoped she wasn't hurt.

Juan, grinning, came near and examined the rope.

"*Caballo . . . nervosa?*"

"*Si.*"

"*Donde esta?*"

I shrugged.

Juan located her tracks, followed them up to the road and

showed me which direction she'd headed. I nodded, but headed back to the truck.

The next development should be clear. The truck's engine was defiantly silent. While I impatiently fiddled with the ignition and accelerator, Juan leaned on the truck door, his grinning face only inches away. My outlook deteriorated considerably.

I pushed by him and started toward the road.

"Lonch!" he called. "*Campo!* Vino!"

Under the sun's blare the chilly dawn had changed to a hot morning. It was nine o'clock when I left Juan and for the next two hours I stalked the dusty road, searching the creek banks and calling "Here, Lark." I changed into a sleeveless shirt. I didn't know if I was even following the right hoofprints anymore.

Sweating, I plunked down on a rock. Yellow jackets swarmed to my naked arms. I took stock of my situation.

My dog and horse were lost. I had no vehicle to help me search for them. The horse, distrustful of people, wasn't apt to wander into any of the scattered cow or sheep camps. If, with the help of friends, I could locate her, she'd be tough to catch. If she chose to head east she could travel to Wyoming and be gone for good. But even the immediate vicinity consisted of hundreds of thousands of acres—the needle in the haystack analogy applied.

Lark had unlimited grazing and mountain springs to drink from. She could live to a happy old age up there. But the dog knew nothing about survival. Probably she wouldn't be able to catch even a mouse, no matter how feral she became. Anyway, her biggest hunger was for attention and human company, and I worried that she must feel desolate.

On top of that, how was I to get home?

I might have at least enjoyed the scenery if immense hunger hadn't gnawed my insides. I felt light-headed. I hadn't had a morsel to eat since the previous afternoon.

It was gloomy to consider that my one hope for deliverance from famine was an amorous herder who'd planned lunch as a romantic interlude.

My appetite subdued my caution, and after a brief rest, I

started the long walk back to Juan's. I guess I figured I'd fill my stomach and be better able to tackle my dilemma. Birds lighted on branches near my head and sang into my ear but I didn't hear them—fantasies of sizzling eggs and bacon crowded my mind.

The climb was strenuous in the hot midday sun. When I reached Juan's he had returned from watering the sheep and was mopping his forehead. I suspect I looked fatigued too, because he suggested a remedy.

"Vinø!" he said. He took my hands in his and patted them sympathetically. He grinned ear to ear. With despair I watched his brawny forearm wrap around my waist.

Juan cocked his head. I did too. I thought I heard a truck.

"*Hombre*. Peekup," I said. It was my turn for an immoderate grin.

Thompson's brown truck appeared on the hill. I sighed and sat down on a hay bale. Juan's smile faded.

"I thought I'd better check how things are going," Thompson said.

"Well . . ." I hesitated. I wanted to be careful what I said because the episode might anger Thompson. He struck me as a serious soul and though he'd told me many funny anecdotes during interviews with him, he related them deadpan. He might be upset with Juan or maybe with me—I feared I'd already become a nuisance.

My pause must have told something because when I glanced Thompson's way I saw his shoulders quivering. I peered around to see he was surpressing a laugh.

"It was a little uneasy," I said.

Thompson threw back his head and laughed. He dropped down on the hay bale and laughed some more. Relieved, I laughed too. Juan joined in. The laughter and a cool breeze that came across the ridge blew away my discouragement. We went in the camp and Juan fixed lunch.

Later, Thompson and a passer-by got my truck running. I felt consoled when Thompson told me he had a band of horses running in the area, and if Lark found them, she'd join them. He gave me the phone numbers of other stockmen to call to alert them that Lark might turn up with their horses.

But when I started down the dusty road for home my spirits slumped. I felt sad to be leaving without Tundra. Her eager personality and sweet disposition were hard to match. I called out the window as I drove and asked picnickers if they'd seen a foolishly friendly gray dog. When I saw the Simmons house ahead I felt worse because I remembered how I'd deserted her that morning.

And there she was, sitting on the porch. When she saw the truck she flew out to the road.

The Simmons family had moved up for the summer that day and found Tundra sitting on the porch. Tundra had charmed the family's girls and they told me, "We were hoping you wouldn't come back."

I thanked them, put Tundra in the truck and endured happy licks all the way home.

If all's well that ends well, I didn't come out of the weekend too badly. I got my pictures and learned things that will be useful to me on my longer trip in July. For one, I will camp a reasonable distance from lonely herders. I had assumed that when it was clear I was there on business all would be businesslike. That assumption didn't figure in the herder's extreme loneliness.

For six days I've fretted over Lark. No one I'd checked with had seen her. Today, Thompson called and said she had turned up with his horse band. She gave him and a helper fits when they caught her, but she's in fine shape, sleek from good grass, and will be coming home in a day or two.

As for Juan, it wasn't his fault I couldn't make my business clear, and his behavior probably was appropriate, given his circumstances. It's understandable he pines for the "girls next door" in the Provinces, the girls I happen to resemble. Anyway, it would take a heart of stone to be angry at ebullient Juan.

For future use, I'm practicing "*Yo estoy escribiendo un libro.*" If that doesn't do the job I've got another one. "*Señor! Yo estoy escribiendo un libro religioso.*"

"Sir! I'm writing a *religious* book."

They call him "Whiskers" but his real name is John Archuleta. He has a long, full white beard which swings above an open cotton shirt in winter or escapes a buttoned-up flannel in summer. His propensity to dress wrong for the season is only part of a larger eccentricity about clothes.

I see him often, walking the downtown streets of Idaho Falls, peering into shop windows, and usually his garb is startling. Sometimes he wears a fringed jacket with western hat and boots and neckerchief, and nearly stops traffic for his resemblance to Buffalo Bill. Sometimes he wears tall argyle socks which meet tan corduroy knickers mid-calf, and above that, a hip-length survival coat. In that outfit he could pass as an elderly Scandinavian who discarded his skis only moments before. Then too, there's the loud red-plaid coat which gives a logger-of-the-Northwest look. Sometimes he wears only T-shirt and ill-fitting trousers, one leg of which catches and hangs on a boot top. In that outfit he looks like a bum.

I followed him through town one morning for several blocks before I got up nerve to catch him and talk with him. He is disappointingly shorter and slighter than he appears at a distance. When I'd driven past him, his beard and apparel made him look wild in a wonderful Old West way, but face to face he seems mild, timid even. His watery eyes aren't fierce like I'd hoped. In his sheepherding days he was a nonconformist and though sheepmen say he was a natural with sheep, dogs and horses, he often lost jobs because he would honor no rules but his own. One time when he was in jail on a drunk charge, he cut a hole in the ceiling and almost escaped. But liquor has doused his fires and now he is merely erratic.

When I told him I'd like to visit with him about sheepherding, he motioned me toward a side-street saloon. I glanced both ways as I entered with him—I hoped no neighbors were around to see me going into a bar at 8:45 A.M. I also pondered the protocol—was I expected to spring for a drink? My purse carried less than a dollar's small change, but I worried that John might incorrectly assume I was on a fat expense account.

Inside, I was shocked to see tables and stools filled at that hour. The dark interior buzzed with talk and jukebox songs. A row of weathered old men cranked their heads our way. John introduced me to several other retired herders. While I was there, a sheepman came in looking for a replacement herder, and one tall old man went to a corner with him to discuss details.

John rambled some stories to me, and it seemed hard for him to keep the conversation's thread. I had a hard time concentrating too, rubbernecking at the leathery faces and trying to imagine what tragedies or intrigues had pushed these men into the solitary sheepherder's life. And why, if they withstood the mountain's isolation, did they need to flock so when they lived in town?

Books about sheepherding I've read sometimes leave the impression that insanity, brought on by isolation, is a common occupational disease. Yet, I've been unable to dig up many stories of actual cases from the sheepmen and herders I've visited.

And R. K. "Bill" Siddoway, a St. Anthony, Idaho, rancher, and former president of National Wool Growers, believes the reverse is true. He told me herders often go to the hills to regain their sanity. Alcoholics, for instance, who want to quit drinking know that to be removed from temptation helps.

Old fellows who can brave the loneliness are a vanishing breed. Sources for foreign-contract herders are drying up too. I always think sheepherding should attract college students, poets or artists because getting away from it all is in vogue. But I've heard several stories about college boys who couldn't endure the realities of herding. Siddoway told me of a young man he hired from the East who burned with love of outdoors and mountains. He came West with elaborate mountain-climbing equipment.

His first disenchantment was watching how much hard physical labor went into the job. Next, he went to the hills, assigned to a flock. After two days he left the sheep and hiked out to a phone booth to plead with Siddoway to pick him up.

The job usually doesn't work out for married men either. Siddoway had one couple, though, Mutt and Molly, who herded for him as a team for years. Mutt once surprised a visitor with the explicit directions he called to his dog up on the hill. "No, no! I said fifty yards to the right. . . . Now go get those black twins by the big rocks." The visitor hung around for a glimpse of the intelligent animal who took directions so well, and finally a head appeared over the rim. It was Molly, who Mutt had sent up the hill. "You've got one back in the trees," Mutt yelled, and again Molly disappeared. Such obedience among women is rare, however.

John pointed to a photograph of himself hanging above the bar. Local artists and photographers are always after him to pose. I understand that; we romantics see something in John beyond the fact that he is colorfully photogenic. He represents the end of an era.

VII

He Leadeth Me in Paths of Righteousness

They found old Joe Carter frozen. In one rigid hand he clutched Dolly's bridle and under the other he hugged his dog. Probably Joe died of a heart attack but no one knew for sure because his body was too frozen to do an autopsy. It was severely cold on the desert that week.

Dolly was tied up at the camp and had a blanket on—it looked like Joe was getting her ready to saddle. The dog was cold but still alive; they pried him free. It was a lonesome way for Joe to go, his friends said, but fitting somehow. His animals were with him at the end.

Joe worked for Shelley, Idaho, rancher Allan Thompson. And one thing that showed up right away in his work was his talent with animals.

One season Thompson gave Carter a horse to break. Carter's camp was parked at the lambing sheds, and Thompson thought Joe might like a spare-time project. The horse hadn't even been handled and she pitched and kicked and fought when Joe tried to lead her. When tied up the horse sat back on her haunches, snorted and tried to toss free. Thompson warned Joe not to do anything foolish like work with her when he was

alone. If the filly hurt him it might be a long time until someone came by the isolated pens.

Some hours later Thompson maneuvered his pickup into the lambing-shed road. Joe, atop the filly, rode up, whistling, to meet him. The green horse wore her bit like she'd always had one on, and Joe wore a smug grin.

Joe did such a good job breaking horses that Thompson gave him one in appreciation. Joe had few material things and it was with some awe that he led off his new horse, a sleek, black, registered Quarter Horse. "Hello, Dolly!" was popular then and Joe named his mare after the tune he heard so often on the radio.

From then on Dolly and Joe were a pair. By the time of Joe's death, they were almost a legend.

Dolly was so ambitious she covered as much territory in a walk as other horses did at a canter. She also possessed a hereditary gift (her sisters had it too), an extra, natural gait. It was an odd run-walk, almost like a pacer, and when she kicked into that gear she gobbled up roads.

She balked at nothing. Joe could head her straight up or straight down the steepest hills. She didn't consider it beyond her abilities to jump wide beaver dams or leap up shale embankments.

But the most distinctive thing about her was her heart. Situations which lathered other horses seemed only to challenge Dolly. She had an inner reserve of nerve and toughness she willingly dipped into whenever Joe asked for it.

Despite her great spirit, she was gentle and calm. They said you could "ride her with a string in her mouth." And they said "she talks to you." Her expressive eyes conveyed many messages—curiosity, trust, humor, even reproach sometimes, if Joe was late with her oats.

But he seldom was. He first took care of his horse, dog and sheep, then himself. If Dolly was an exceptional horse, Joe was equally exceptional as a trainer. When he held an animal's face to talk to it, onlookers sensed something magical. His dogs worshiped him, his sheep clustered around him.

As long as Joe could remember he'd always liked animals. But he hadn't always been handy with horses—he'd had to learn to trust them. When he first started herding as a young man he faced perilous mountain paths with apprehension.

Slippery, narrow footing was common where he trailed sheep and a misstep could have been fatal for rider and horse. He peered over his horse's shoulder into canyons hundreds of feet deep. On sheer naked cliffs not even a weed would slow his fall if a horse slipped, and he found it hard to surrender himself to his horse's care.

His temptation was to steer the horse where he thought the path was safest. Or get by treacherous places on foot, leading his horse. But seasoned riders convinced him he should allow the horse to choose its own best path. And he was more apt to be hurt afoot if a horse slid down on him from behind or jumped onto a ledge where he stood.

One spring he was trailing sheep on the "W" south of Taylor Mountain in southeastern Idaho. The path is rocky and steep but that year it was especially bad because of late deep snow. The boss went ahead and shoveled a narrow path up the mountain. He hollered to Joe it was okay to come on, but warned him it would be a mistake to hesitate anywhere—"keep 'er movin'," he ordered.

On a steep stretch Joe got scared and reined up. The horse, who had been moving along well, started to slip as soon as it hesitated, and in a few moments it lost its footing and tumbled. Joe was unhurt and the horse only got skinned up, but Joe learned a lesson that day.

Joe became as competent a horseman as anyone could mention. Dolly was his last horse—he had her nine years. For all that time she was his alarm clock—each day at 5 A.M. she banged her head against the sheep-camp door wanting her oat breakfast.

He never tied her. It worried Thompson that even in the desert, where a man must have a horse to ride to get water, Joe would allow Dolly to wander three or four miles away, confident she would return in the evening.

Sheep-camp horses generally ground-tie quite well. When the rider gets off he puts one or both reins on the ground and the horse stands still as though tied. Joe didn't bother with that ritual—he hopped on and off at will with reins in place and Dolly never budged.

Thompson remembers how aggravated he'd get because Joe wouldn't put his oats out of Dolly's reach. Horses will overeat on rich feed if they get a chance, a common cause of death.

"I'd warn Joe, 'You're going to lose that horse if you're not careful, she'll get into those oats and get sick,'" Thompson recalls. "But Joe would shrug and say, 'Naw, she won't. I told 'er she's not supposed to.' And she never did. There'd be an open grain sack, and she'd walk all around it and not get into it."

Other men sometimes felt annoyed at the way Joe pampered Dolly on the trail. Once Thompson noticed a rift between Joe and another herder, J. C. Mondragon. Joe and J.C. were packed out together, J.C. was cooking and Joe was tending the sheep. The two men worked well together, ate meals together and slept side by side in the teepee. But they'd stopped speaking. Thompson tried to get to the bottom of the problem, but neither man would give a hint, and the silence went on for weeks. Finally Thompson got Joe alone and made him talk.

"Hell, Allan," Joe said, "all I did was let Dolly drink outa our water bucket and ole J.C. got mad!"

But men appreciated Joe's good humor too. Once on the Caribou Basin Trail mosquitoes were so bad the dogs wouldn't even work, they hid in the creek. Carter was on Dolly and leading another horse. Mosquitoes sat side by side on the horses' necks. Joe leaned over and asked the camp tender in a sincere voice, "What are we going to do for horses when you take these to town? I'm sending 'em in for transfusions."

Another time he commented on the horse he was using while Dolly's sore leg healed.

"He's a good horse to have on a fence-buildin' crew. He finds every hole in the county."

His own horse, he bragged, could streak across an unfamiliar field at a hard gallop and never trip in the numerous badger and rodent holes.

Joe had arthritis and it got worse with time. Once when Thompson drove out to the desert to deliver groceries Joe came riding up unsmiling. He told Thompson he'd been up on a ridge during a storm, praying for lightning to hit him because the pain was so bad.

Feed was in short supply and Thompson needed to start feeding pellets. They came in 100-pound sacks and Thompson asked Joe if he could handle it. Joe said he thought he and Dolly could.

First Joe divided the feed into fifty-pound piles and resacked it. Then he piled the sacks on the wagon's fender. By this time Dolly knew she had to scoot over next to the wagon for Joe to mount because his arthritis didn't allow him to use the stirrup. She side-stepped next to the wagon and Joe dropped into the saddle. He took her out and showed her a giant circular path they needed to take when he distributed feed—his hands would be full so he couldn't guide her.

Then he went back and picked up sack after sack off the wagon's fender. Dolly remembered where she was to go, and day in, day out, that's how the two fed 1,800 sheep. That might have been where they were going when Joe's heart attack struck.

After Joe's death Thompson never put Dolly with another man permanently. Sixteen years old now, she runs with a horse band in the Willow Springs area.

But whenever she does go out to help trail sheep other herders ask with a certain respect, "That Dolly?" And if her rider is a man who usually mutters and cusses at his horse, he'll hold his tongue. Men ride up and stroke her head and they say, "Good ole horse." And they say, "She's more horse at sixteen than most horses ever were." And if there are oats around someone goes off to get a scoop while someone else unbridles her.

It might be that the horse, with her sleek black hair and knowing eyes, commands respect. Or it might be respect for the memory of old Joe.

I now spend a day every week in the hills, and whenever I can, I take the boys. When the day involves horseback trips or serious interviewing or hectic scheduling I can't manage it. But if the day looks leisurely enough to accommodate the supervision of two vigorous three-year-olds, the boys go along.

They almost explode with excitement when they learn they're going to the mountains. A few times I've taken along books to read to them if they get bored, but when we arrive in the hills I see what a foolish, town dweller's notion that was. As if the mountains didn't contain endless amusements.

If you're a little person, pine cones are a favorite fascination. You collect them, throw them, smell them, hide them, turn them over and over in your palm for inspection, pull apart the pods and finally give the bare cob to your mother as a present.

Badger and rodent holes are first-class entertainment. You can peer into their darkness, speculate about who lives inside, holler greetings and announce your name, invite the critter out to share peanut butter, probe the hole with a long stick and ask the tall people questions they don't know how to answer. "Does he get wet in here when it rains?" "How does he shut the door? His mamma doesn't want the flies in!"

The sheep camps are a wonder too. Because they are miniature, they seem superior places to live.

Narrow steps, to be climbed and descended fourteen times, divert you several minutes before you decide to enter. Once inside, you can rub the wood stove to gain a wonderful black stripe in your hand, which can quickly be shared with your face. If you pet yourself against the stove like a kitty on a table leg, you can produce a great smudge which starts at your ear, runs the length of your shirt and ends on the hip of your clean jeans. Sometimes, of course, the stove is fiery hot and presents a hazard to handle with big-boy maturity. "Be weally careful, honey," you caution your brother.

It's an easy scramble to get on the bed, and once there you can peer out the small window onto nearby ridges. Your own room should have such a window, you observe, so you could watch squirrels in the trees. The fact that your backyard has no such trees doesn't concern you.

But the very best thing, if you're little, is the board that mysteriously folds out of the wall to become a kitchen table. Some nice treats appear on that table, but they can't tempt you away from the more fascinating work of pushing the table back up, hooking it, unhooking, bringing it down, putting it back, etc. And when your mother wonders if that is all right a herder comes to your defense and says sure, let the kid enjoy himself. That's the nice thing about these wee houses, you can relax in them. Spills and other *faux pas* pass unnoticed, unlike in the big houses in town.

I like the herders even better for the kind way they treat my children. The men seem happy to take them by the hand and lead them on walks, pointing out things of interest. Language is no barrier here—when the herder spies something he wants to show the boys his faster walk conveys excitement, even if his words are unfamiliar. It works in reverse too, wide-eyed boys have no trouble transmitting their gratitude about some rare find.

Usually these treasures go home with us for display in their bedroom. So far we have two sheep skulls, half an antler, a smashed tin thing, thousands of pine cones and a fragile white item I think is petrified dung.

One herder, Bill, took Matt with him for a rather fast horseback ride. Matt enjoys riding with me, sitting in front of me in the saddle, but I've always taken him at a sedate pace. Bill chased fugitive sheep while Mattie, hanging on as the horse reeled and turned, had a wonderful time. To be impartial Bill repeated the ride with Shane. By the poor little guy's panicked face I'd guess he had a miserable ten minutes.

Herders search their camps for treats, and when they find a package of candy or cookies, they press the boys to have enormous helpings. I've protested when I was sure the boys had had enough, without result. I hated to be too stern with the herders, but my boys are unaccustomed to sweets. Consequently bellyaches and tears have plagued our return trips.

We were in the hills one day to watch sheep load onto the trucks. One exceptional dog was keeping the lambs moving. When the lambs slowed the dog would leap onto one's back and ride it toward the loading chute. A herder I'd never met, noticing that the boys were eye level with fence boards and could see nothing, strode over and picked them up. When the boys could watch the activity it brought peals of glee. "Dog widing lambie," Shane squealed. The herder thought it was worth some howling laughter too.

It takes so little to gratify three-year-olds. In that way they have more in common with the herders than they have with other grownups. Unaccustomed to artificial amusements, the herder sees charm in his simple surroundings. He has no marvelous creations of man with which to impress guests, so he shares the routine magic of nature. A child is a rapt audience.

I envy the herders' simplicity. Unconcerned with vehicles, homes and appliances, free of social obligations, complicated scheduling, self-improvement regimens, scheming and competition, they can have habitually uncluttered minds. And an uncluttered mind can suck up what their primitive life offers—scenery, free time and enhanced awareness of nature.

Herding, it seems to me, allows grown men to repeat some aspects of childhood. Their life largely shelters them from the complicated "out there" world. Animals are significant in their world. They have time to weave weeds into odd shapes and gaze at empty skies.

Some herders are so keen. They live in quiet, so their unjaded ears can detect small rustles in the weeds. They've trained their eyes to identify faraway animals as friend or foe. I saw one stop on a path, sniff the air, then trudge off through the grass to find a favorite flower.

Is this generally true? Or is my vision selective and I choose not to see grumbling boredom with flocks and disregard for views and vistas?

I'm inclined to think the occupation itself weeds out those men who find the life disagreeable. The man who requires fast action, artificial amusements and bright lights couldn't bear up for long.

And I have evidence of how reverently some herders feel about their surroundings. Erosion so distressed one old Mexican that he used his free time to patch the damage. Where he found erosion he took up his shovel to build an earthen dam. He worked so painstakingly that his dams still endure. The illiterate old man knew nothing from books about erosion's long-term effect—he only knew that the land had been disturbed and he wanted to set it right.

Legends abound about herders and their perceptiveness. Herders' family members tell how Dad or Uncle Jack or Uncle Fernando could predict winds, storms and catastrophes.

"He would watch the ants, and know if a thunderstorm was coming," or "He knew he was going to die; the animals behaved differently to him."

Herders' pleasures, too, are simple ones. Some herders take

up knitting, others wood carving. One Idaho herder carved necklaces and ornaments of wood, and presented them at random to visitors.

The experts count and measure natural wonders and natural routines from a distance. The herder participates in them. He knows nature from the inside, and has a spiritual kinship with creation.

I haven't yet taken the boys over to Lennie's camp. I think they'd enjoy seeing a child who lives in the camps. Big Lennie, who is divorced, and little Lennie, eleven, tend their sheep in a spectacular location. Their camp is on a ridge beneath a beautiful range of peaks. At night they've been sleeping in a teepee on the mountainside, near the sheep, because of coyotes.

Each day they return to cook a pot of delicious-smelling beans, then at twilight head back for the hillside. It created a memorable scene the night I watched them, in fading daylight, climb a faraway hill. Little Lennie, his arms around his father's waist, rode behind the saddle on the colorful pinto, and just beyond, their white teepee flapped in the wind.

Lennie thrives on sheep-camp life, if his bright little face is any indication. He is eager to answer questions about sheep care, tell about the new late lamb in their flock and about how he and his dad huddled in the tent the previous night while lightning crashed overhead and wind battered their canvas shelter.

He has questions of his own—how much did the camera cost, does it have different, you know, lenses things, do the pictures always come out? When I put the camera strap over his neck he directed me where to pose beside the sheep camp. The camera hung askew and pointed at the ground, but Lennie seemed to be having a good time.

I envy him a little, and wish I could either repeat my childhood up here or raise my own boys in a sheep camp.

VIII

Yea, Though I Walk Through the Valley of the Shadow of Death

Name's Jess.

Uh guess uh been at it 'bout slong as anybody—been herdin' fifty years.

Yeah and uh worry. Uh guess everbody does. Whut uh worry 'bout is gittin' dumped. If you was ta git throwed in eight er twelve inches a snow un broke yer laig, it'd be too late by the time they found ya. That's why ya gotta have a good horse under ya that's got no foolishness in it.

Uh guess a herder lives with death more 'n most folks. What uh mean it's all around, jis part a nature, but that don't mean he's lookin' forward to it hisself, WHOO-OOOOOOO!

In the ole days when uh started un they pulled with wagons, two men was tagether, the camp tender and the herder. Ya din't worry sa much then. Now someone is s'posed ta check on ya ever seventy-two hours, but gittin' hurt out here is somethin' that bothers everone.

Slow down a little bit, Bob! Durn dog. Hey, them sheep are awright, Mike! Come back here you two! They're doin' fine without yer help.

When uh was younger uh used ta break muh horses right up here. But when uh hit now it takes longer ta git up. That's what we worry 'bout most, all of us, gittin' hurt and not bein' found in time.

Kinda funny how those rocks stand on end there, like a natchurl fence.

Bob! Mike! Slow down, you'll last longer! Them sheep ur goin' fine! Them dogs has been on the farm all winter and they're na good.

Un sure, there's other stuff ta worry 'bout too, besides fallin' off yer horse. If ya git sick yer a long ways out.

One time uh had tick fever. Boy, I'll tell ya uh was sick, that's right. Woke up one mornin' in muh teepee and found one awready dug in, by muh groin. That's a mighty bad place ya know, cause of a lotta nerve endins. Uh pulled it out with muh bare hands, jis like uh awwus done, but maybe some got left in. Uh was in the hospital fer a week. They give me drugs un medicines but mostly they jis changed the sheets. Uh soaked a set of sheets 'bout ever thirty minutes uh was so hot.

It took me a month a bed rest ta git right. The doctors said it weakened muh heart muscles, sorta like whroomatic fever does. 'Bout that time uh tried ta git insurance, but with muh heart uh couldn't. Few years later though the ole doc said uh'd whupped it—probly wasn't twenty-five men around as healthy.

Say, ya never git too fat with this life. Ole Dan Fisher come up ta me at the store in Bone this spring and says, "Good hell, Jess! Don't drink no tomata juice. Folks ud take ya fer a thermometer!"

Now take ole Dan, he's a good man, but he didn't work out as a herder. Gits too nervous. A sheepherder don't git nervous, jis takes things as they come.

Uh used ta quote 'im that little pome, "If there's a task fer you ta do, Put it off a day er two."

Uh never seen a tick on a dog—don't know wy, they jis don't seem ta git on 'em. Ya kin tell a wood tick from a sheep tick. A wood tick has a brown body and looks like a bedbug. A sheep tick is gray-bodied like them ole monkey-headed spiders, the body in two parts.

Bob! Mike! Godawmighty you two take a lotta yellin', a man could make hisself deef. Ya got some lambs down in the wash! When they're tagether they ain't na good. One or the other is awright, but ya git em tagether un they jis wanna play.

One time uh come up on a sheep camp and it smelled like smoke. Uh opened the door un there sat the herder like this, on the bed, frozen-like. His hands was curled up like claws un ya couldn't make out 'is head from 'is shoulders—he was jis one big swolled-up blister from 'is chest up.

He'd been fryin' bacon on the cookstove. Some grease splattered inta some oil un the place blew up in flames. He couldn't git ta the door. Lucky thing, the fire went out by itself.

But that's how uh found 'im, he couldn't move. Lucky thing muh boss was along un we got 'im out. The boss took 'im ta town in the truck un oh God, don't ya 'spose that was a painful fifty miles? Then they got 'im inta a ambulance ta take 'im the rest a the way ta the hospital. They had ta make a opnin' in 'is throat because 'is throat had been burned, but the man lived. Uh think he's alive ta this day yet.

Mike! Bob! Those two! Din't have na discipline as pups un they're no damn good. Stay away from them sheep!

Muh fren John fell off a hay wagon once. This was in the ole days when we fed loose hay offa the wagons, with horses pullin'. John took a fall onta a patch a ice and it sent 'im sprawlin'. It got 'is tailbone un he flat couldn't move un he was in pain. So uh took 'im down ta the hospital un they x-rayed 'im un sure nuff it showed up he'd hurt 'is tailbone. But there was somethin' else on them x rays un the doctors couldn't figger it out, jis somethin' unexplainable.

We was all lookin' at the x ray and uh says ta the doc, "Ya know ole John wears long handles? Did ya think it might be the button on 'is trap door?" Whooooooo! Un sure nuff that's what it was.

Now let me show ya somethin' here. This here is wild carrots. See them little things on top as long as the tip a yer finger? Herders takes 'em un throws 'em in stews and soup and it flavors it jis like carrots. Taste 'em. Most folks thinks they taste jis like carrots but I think they tastes like anise. But when ye throw 'em in soup they tastes like carrots.

People gits hurt movin' camp too. Uh remember a camp tender who come a-barrelin' down a canyon, had too much ta drink uh reckon, goin' way too fast. He rolled that pickup inta the willers by a crik. A by-passer come up un seen the truck upper side down un tied ta git 'im out but it was too late, he awready was killed.

There's a cactus blossom fer ya Debbie. This horse ud fall off a trail ta eat one a them purple cactus flowers. She's a right good horse though, a real lady, the kind ya like up here.

Bears kin be a danger too, but bears and coyotes aren't much ta attack humans un they only would ta perteck their young. I've never knowed one to go off their rockers.

Muh fren ole Pete, he's eighty-two years old un straight as a string un tougher 'n a goat. About five years ago he was havin' trouble with the bears where he was herdin'. They'd been gittin inta is sheep and tearin' 'em up. One day 'is dog chased a sow un two cubs outa the sheep un Pete follered 'em. He's a dead shot un he killed the two cubs un they fell outa the trees. He hit the mother right twixt the eyes but she didn't fall un he knowed he'd only wounded 'er.

He dressed out one a the cubs un then went ta look for 'er. Here's the part that scared me. The underbrush was so thick he had ta crawl on 'is belly un uh'm shore glad he saw 'er before she saw 'im. She was within five feet a 'im when he shot 'er again un hit 'er right through the lung, finished 'er off.

Uh've had 'em kill up ta nine sheep at a time. Bear won't eat fresh lamb—oh, they'll eat the bags on a ewe, un the liver un the lungs, but then they waits fer the meat ta rot and the maggots ta git in it. They like rotten meat.

Another time a guy was herdin' up by Palisades un he'd been losin' sheep ta the bears. He come up on one in the sheep un started ta shoot it, but he shot 'isself by accident.

When uh packed 'is groceries in uh knowed somethin' was

wrong, the dogs was gone, they'd stayed with 'im. He tried ta drag 'isself away ta find help instead a stayin' at the camp. Uh tracked 'im, but it wasn't till two days later uh found 'im. Uh brought 'im back on horseback, laid over a pack saddle. Un he lived too, yes sir, un that outfit is still payin' 'is wages.

Uh've been ta some sad sheepherder's funerals. Ole John Johnson died this year, hit by a car. They said the radio told he stepped offa the curb inta the street at two in the mornin'. Guess he'd had too much ta drink. Un then a month later they found ole Ben Decker dead. Guess that quick newmonia got 'im, seems like a lot of 'em gits that. Uh think it comes from drinkin' too much un layin' out in the cold.

Uh made it down ta Ben's funeral. Me un another man was the only ones there. Kinda sad.

But the worst funeral uh was ever at was fer Cliff Hames. In the ole days it was a little diffrunt. When ya went out ya ate un lived un slept with another man fer months at a time. Ya go ta know 'em real good un ya got attached, even though ya'd git mad at 'em.

Uh knowed Cliff like that—we'd packed out tagether many times. When he died the county burried 'im. Uh don't know wy they was in such a hurry, he was s'posed ta have people someplace but they didn't even try ta git aholt of 'em.

Cliff was a big feller, big shoulders, lotta muscles. When the county put 'im in a pine box he was too big ta fit. They couldn't git that box nailed down all around un uh'll tell ya what, that was hard ta see.

Lotta these guys starts stayin' in town when they're old enough fer social security. That social security has ruint many a good man. There's men that do fine out here and stays healthy, but when they start stayin' in town they go downhill right now.

Uh remember ole Loren Jones, when uh was jis a kid he tole me, "Schools un unions un politicians, that'll be the ruination of this country," un uh'll tell ya what, he wasn't a-missin' it too fer. He was in his sixties then un he said, "I won't live ta see it but *you could*," jis like that. If there ever was a racket un a nasty one it's these politicians.

Mebbe if a man has hisself a wife he don't end up sa lonely. Uh never married. Uh was engaged once. Name was Delores. She come ta see me ever weekend up at the camp. She was a-comin' on the Skyline Road one time un a car hit er head-on. Killed er.

That there yonder is the Fourth of Jewly Ridge. Did anyone tell ya how it got its name? One time on a Fourth of Jewly weekend a bunch a sheepherders got tagether on that mountain. They had a helluva drunk. Next day the sheep from a buncha bands was all mixed tagether. It was a helluva mess. They had ta build a corral up there ta sort 'em all out, un it took 'em days, un that's what they named 'er after.

What ur them fool dogs doin'? By hell, they're after that blind ewe. BOB! MIKE! Uh bet a bullet'll stop 'em. *Un look here, uh don't even have a shell in muh gun. BOB! MIKE!* Right off through the underbrush there, them no good sonsabitches! Uh guess they'll run 'er ta death.

Uh didn't nurse 'er along this whole trail ta have no-good dogs take after 'er! If ud had a shell uh would a stopped one a 'em, then mebbe ud have *one* ta work *right!*

Boy, uh'm ready fer a cup a coffee. When uh git muh fire goin' that's the first thing uh'll unpack. Coffee in town don't taste like it does up here—uh've had some good coffee in the camps.

An' there was a funny thing that happened once, and uh've heard of it happenin' ta others. A bunch of us got tagether in Brockman, we hadn't seen each other fer a year. And we got ta drinkin' coffee, cup after cup, no food. Finally uh rode back ta muh camp, had more coffee while uh fixed dinner, and ya know, when uh went to stand up uh couldn't move. Uh mean, uh flat couldn't move. Uh think it's the kidneys that does it, uh've heard of other people gittin' it. Uh fell asleep right in front of muh camp and next mornin' uh was okay, but that scares a feller awright.

Here come one a them dogs. He's got blood on 'im. Blood all over 'is head. Uh bet they killed 'er.

Git over here, Bob! Uh'll whup im *good.*

Moan you no good S.O.B.! Yer lucky it's jis a kick un uh don't wring yer damn neck off. That Mike won't even come close. Now Bob's slunk off too. Uh got no dogs. Oh well, the sheep won't know it today. Them ole ewes know where they're goin', so uh'm awright fer today.

Ya gotta train 'em as pups. If they get no discipline they'll do that sometimes, go bad like that, but uh can't remember this happenin' fer a long time.

Here she is. Ain't that a shame? . . . She's purty too. She'd been a-doin' fine on the trail, but them fool dogs scared 'er and she bolted off un they jis run 'er and run 'er and bit at 'er and killed 'er. It's a shame. A little bit discouragin' too. She's the buzzards' now.

Sometimes uh don't know wy uh do this year after year. But uh guess sheepherdin' gits in yer hair.

Well look what we come onto here. Ain't that somethin' . . . a whole field a columbines hid up here. Look at the size of 'em, as big as a woman's fist. Uh like them creamy petals and pale yella middles.

Maybe that's what we come back fer? Fer comin' up on a crik an findin' a big patch a columbines hid up. Uh think uh'll take a bookay along fer muh camp tanight. There's really nothin' much purtier 'n patch a columbines that surprises ya when yer jis goin' ta git a drink.

Yep, there's worries up here. Un hardships. Un death all around. But the columbines brings us back.

IX

I Will Fear No Evil, for Thou Art with Me

There is much to fear when herding.

The herder has to fear for his sheep and for himself.

But, says Nick Padilla, "if you're scared up here you can't do nothin'.

"Some people worry about bears, rattlers, hornets and people. But I walk all around up here. Once a man lets himself get scared he's all through, he might as well go to town."

On the Caribou National Forest a few years ago a bear slashed a teepee. The herder wasn't home at the time. When he returned and found the shredded canvas he couldn't stifle a fear that the bear would return. His camp tender assured him bears seldom trouble people and that a rifle shot would frighten away any bear. But the herder demanded a replacement as soon as possible and friends think he left herding altogether.

One man who tracked a cat that was troubling his sheep clambered onto a rock ledge and found himself face to face with a cougar. Something held the man spellbound and the cat turned and fled.

A man accustomed to the hills can withstand small aggravations as well as real menaces. In late summer swarms of yellow jackets and hornets infest the camps and are everywhere indoors. Men get so used to them they even allow the insects to perch on exposed skin. Many men say they never get stung despite season after season among the touchy creatures.

Lightning is a hazard to the herder and his flock. Most old herders can give accounts of a comrade who perished on a high ridge during a thunderstorm. Finding good cover on a naked hill is a problem for a man on horseback—he can't desert the sheep to seek lower ground.

John Hunt told of a tragedy that killed three hundred sheep, two herders and three dogs. The men and animals were near some big trees, waiting out a sudden storm. One dog who survived wandered into a nearby camp next day, bewildered.

Allan Thompson once had eighty-nine sheep killed by lightning. The animals were bedded down close together. The bodies didn't deteriorate for some time and it "seemed eerie up there. I almost hated to go up and look at them," Thompson said.

Accidents with firearms are not unusual. John Hunt once shot off a toe while tracking coyotes in Oregon. He fell while crossing some logs, and the rifle fired.

Jim Mays told of a recent accident which involved his herder, Tony.

"About 3 A.M. Tony heard a bunch of racket down at the sheep. He piled out of bed thinking 'Bear, bear!' He lit his lantern, put his clothes on, saddled his horse and took off.

"When he got down there everything seemed to be in order. Next day he found elk tracks, so evidently elk going through had spooked his sheep.

"He started back to camp and at that time he realized he had his house slippers on, and he still had that mountain to climb to get back to camp. His toes started to blister, so he got on his horse.

"We have this propane cannon to scare coyotes that lets out a big *BOOM* every hour. Tony was almost to the rim where his camp was parked and that dang thing went off.

"Tony had tied his rifle up close in front so it wouldn't catch on a tree. When his horse rared, the rifle came up and hit him in the face, knocked him out and he fell on the ground.

"To hear him explain it, he came to saying, 'Am I dead, am I dead?' All of a sudden he felt something warm in his mouth and he started spitting blood and teeth. The blow shattered his dentures.

"He managed to get his bearings. The horse had gone off through the trees. He got himself to camp, cleaned up and hung a lantern in the trees so the horse could see it.

"He had another horse there, he tied it near the light, and pretty soon the two horses were whinnying back and forth. In a half hour the horse came back on his own.

"The horses were used to the cannon, but it just happened it went off pointed right at Tony and his horse."

Many herders go years and years without a mishap on horseback, but when bad luck strikes, a man can end up with serious injuries.

One year Virg Shinn was starting sheep into a lane when a sheep tripped his horse. He was a little downhill and the sheep "darn near run over me."

He was riding along a canyon rim when he was sixty-two and his horse slipped over the side. The fall broke every rib on one side of Shinn's body and punctured a lung.

"The old doc told me, 'Don't never break no more ribs,'" Shinn said. But this year, at age seventy-five, Shinn had a horse go over backward with him and he broke ribs again.

"Now the doc says, 'Don't never get pneumonia. I don't know what's protecting your lungs now—your ribs are in chunks.'"

Long-time herders can remember when cowboys were a principal threat. In the days of animosity between cowmen and sheepmen, isolated herders were easy targets for abuse from cowboys. The herder often was unmounted, unarmed and spoke little English.

Nick Padilla remembers an encounter he had as a boy.

"I was out herding with my brother. I was just a little kid. My big brother, he had only one leg. Here came riding up a big, stout cowboy. You could tell by his look he was going to do something.

"And you know what he did? He roped the burro, roped it and dragged it. I was kinda shaky awright. My brother and I ran and hid in the trees.

"We didn't know what to do, but we had to do *something*. The sheep were out there. My brother picked up a rock and threw it. It hit the man in the head. I picked up one and it hit him on the jaw. He rode away. We were plenty scared he would come back, but we never saw him again."

Joe Aldana, Idaho Falls, Idaho, a retired businessman, worked in this country as a herder when he arrived from the Basque provinces in 1918. At that time rumors were rife about cowboy atrocities.

"In those days we trailed sheep three hundred miles from Soda Springs to the Nevada desert. I heard of many instances where cowboys tried to scare away the poor herder. I even heard that cowboys had dragged one herder to death, but I never saw any of this trouble myself."

J. C. Mondragon, now in his nineties, remembers the days when every herder and cowboy in a saloon kept his hand near his gun in case of trouble.

"The cowboys thought they had it over the herders because they lived in bunkhouses and the herders lived in teepees. What they didn't know was how much easier it is to herd sheep than cows. Sheep stay together," Mondragon said.

Along with the fear of actual dangers is an undefined concern that protracted solitude may damage sanity. In the Basque provinces people use the expressions "sheeped" or "sagebrushed" to describe the condition of some herders who come home mentally broken.

"Ya worry sometimes, when ya think somethin' is real but ya suspect it might be a vision," one old-timer said.

Scattered rumors go around of irritability that led to shootings, depression that led to suicide and desolation that led to crying jags. Yet, herders seem to have little firsthand experience with such events. Drinking bouts are by far the most common indication of problems.

"Yes, I have heard of those who became mentally unbalanced," Joe Aldana said. "But whether sheepherding did it, who knows? Maybe they were going that way anyway."

For foreign herders, fear sometimes springs from being in a strange land among strange people.

One recent spring about a hundred men organized a search party for a missing herder. The herder, a Basque, worked for sheepman Jerald Stolworthy. Stolworthy found the man missing from the camp, which was parked on the desert amid lava outcroppings.

The man had taken a walk near evening, and it was dark when he started back to camp. He lost his way, and couldn't get back.

Temperature dipped to twenty below zero that night. The next day searchers picked their way between the rocks, peering apprehensively into deep chasms cut by ancient volcanic spills.

The man turned up safe two days later. After wandering awhile he burrowed into a farmer's haystack to keep warm. Lights from the farmhouse shone only a short distance away, but not knowing any English or any Americans besides his boss, he was afraid to approach the house.

Sometimes herders must act boldly to save sheep. Mays recalls two episodes in recent years where enormous losses from bears would have been even greater without the conscientious efforts of his Basque herders.

One year, on Big Elk Mountain, a bear stampeded the sheep at night and the frightened animals piled up on an area near the bed ground. The area wasn't very steep, but when the herder reached the sheep many were trapped, though still alive. Working in the dark, the herder lifted about a hundred squirming bodies out of the pile-up. It was almost shipping time and lambs weighed about 100 pounds, ewes about 180. The herder hauled up sheep until he was too weary to lift any more.

When Mays and the herder went through the dead the next day they found only fifteen or twenty with bear claws on them or with broken necks. The others, almost 150 head, died of suffocation.

Mays mourned the loss but figured it was a freak once-in-a-lifetime thing. The next year, however, the situation was repeated, with even worse consequences.

This time the herder got to the sheep in time to see three bears, a mother and cubs, he thought, fleeing. Fatality count the next day was 230 head. Again the loss would have been much higher if the herder hadn't worked through the night to lift about a hundred sheep out of the morass.

There is much to fear when herding.

But if a man gets scared he's all through, he might as well go to town.

X

Thy Rod and Thy Staff, They Comfort Me

When I was pitching my tent last week a camp tender came by and told me a bear story. He said he shot a vicious bear on a nearby ridge two years ago. He wore a big grin as he told the story, so he may have been trying to scare me.

He didn't succeed. I'm not afraid of bears or cougars or coyotes or Big Foot or human intruders. Just field mice.

They've started to invade my tent in my absence and the other night before I crawled in my sleeping bag I noticed movement under the top brown cover. I carried the bag outside and shook out a mouse. I shook it a thousand times more to make sure the mouse had no companions, but still I worried during the night.

So I've moved my food outside and deliberately left the apple sack, which the rodents already had started on, near a tree. I figured if they were busy filling up on apples they wouldn't be tempted to search my tent for other snacks while I slept. They enjoyed what I left them—I found several apples gnawed away this morning, and even the paper sack had great holes in it. I think my apple supply is large enough to divert the mice two more nights. Then I will go home.

I've been here six days. My tent sits in a grassy spot near some trees, well back from the gravel Shirley Creek Road. A fence runs along the road because this place is used for summer grazing. Pasture extends some distance west, and just east of the tent is a quaking aspen grove where I tie the horse. Beyond the trees a hill plunges down to a swamp and a river and a bubbling hot springs.

Six miles east is the Brockman Forest Service Guard Station. At Brockman sheep outfits are going on summer range this week. Rangers count the sheep before letting the ranchers move on to summer allotments, so it's a good place to hang around about 6 A.M. and visit with ranchers and herders.

Life is very pleasant here. I thought, never having been away from my little boys, I might worry over them constantly. Their father took them camping at Glacier Park, so I assume they are well occupied. Still, when they pulled out of the driveway waving their little arms, I cried.

During my first night here I awoke several times to the friendly sound of rain thumping my canvas roof. It has been several years since I've passed a night all alone, and I realized I was enjoying it. Then yesterday someone mentioned the date was July 7. I remembered July 7 is my boys' birthday and it surprised me how little they've been on my mind.

The busyness may be a factor. I forgot my watch (I'm glad) but by the sun I guess it is 5:00 or 5:30 A.M. when Tundra and I get up. If I were tempted to sleep in, I couldn't. Compared to the quiet country where I live, this place throbs with noise. Woodpeckers, quails and other birds start a racket at the first hint of daylight. Sandhill cranes, with their odd, jungley call, sound close, but I've seen their residence and it's a good mile away. In the distance, cows moo.

After I pull on my sweat shirt and jeans and step from my tent I hear another sound—this one ominous. Jets in formation. Sure enough, I'm their target and I try to outrun six million mosquitoes on the attack.

Unfortunately a couple of times each day I must bare a large area of skin and mosquitoes are great to take advantage. If I'm unable to sit down when I get home it won't be from saddle sores. The first couple of days I thought I wouldn't be able to stand all these bugs, but it's funny, I'm noticing them less. Judging by the number of welts I have, they may all be full now. Anyway, by midmorning they are gone, except in the trees.

The sun first lights the Skyline Ridge to the east. Sugarloaf Mountain and the Outlet Canyon behind my tent take on a hazy glow. Up here I can best appreciate what an old friend of mine used to say, "Many sunsets are nice, but every sunrise is beautiful."

I go right up to Brockman and find herders, ranchers and Forest Service personnel already there. When I explain my project sheepman ask, "Are you going to tell about the coyotes?"

Despite the relaxed cordiality, I think the Forest Service and the sheepmen are uneasy with each other. Rangers joke that the Forest Service doesn't want to be blamed for the sheepmen's woes, and one sheepman confessed he thought at first I had some connection with the government.

Tundra's behavior has taken an odd turn up here. At home she never goes trail riding with me. She follows to the edge of the field, then sits down and watches me ride away. I guess she's decided it's no fun to trot beside a horse under the desert sun. Even when I ride my bike she flops down beside the pasture and waits there for me to return. At times I've felt disappointed that she isn't more of a working breed and doesn't hold up well on outings.

But the mountain air has pumped unlimited energy into her. She runs from dawn until night, bounding through the bushes and darting under stones after chiseler squirrels. Each time I look up she is racing after something—an animal, an insect, or a bird. One day we joined a sheep drive, and by day's end she was even chasing the sheep a little. The herder kidded that I should leave her up here to become a sheepdog.

At Brockman the other morning she did what was for her an astonishing thing. While I visited with a ranger she took up with an outfit moving a band of sheep. The lure of tinkling sheep bells, crisp air and other dogs overcame her usual devotion to me.

When I discovered her missing I drove after the sheep band.

From a distance I could see her gray coat bouncing above the weeds. She refused to come when I called her and I had to wade through knee-high wet grass to capture her. She was muddy and squirmy as I carried her back to the car. The car reeked as we drove off—with her pining out the rear window and me incredulous about her infidelity.

I'd assume that after such adventurous days she'd be exhausted by evening, but if I decide to take a twilight ride her eyes light up, her ears perk. When I open the gate she whirls in circles, jumps against the horse's chest, then dashes off in pursuit of something or nothing.

At bedtime she flops down near my sleeping bag, but during the night I wake to find paws across my chest and a moist nose in my ear. I wanted to make her sleep outdoors—I worry about the ticks she is collecting in the weeds and ticks aren't something to be complacent about. I've seen two wood ticks stalk across the floor, stowaways on Tundra's thick fur I suspected. But a long-time herder told me wood ticks don't get on dogs here. That doesn't square with what a veterinarian told me, but I'm persuading myself to believe it because Tundra would be insulted to be put outdoors.

I return from Brockman about 10 A.M. to fix breakfast, usually a cheese omelet or scrambled eggs with vegetables. I've built a couple of wood fires to try pine pitch, which lives up to its reputation for getting fires going quickly. Still, it's usually faster to fire up the one-burner fuel stove.

By now the bugs are gone and I sit on a rock and chew my breakfast with exaggerated leisure while taking in the view. While eating I heat water for dishes and a bath.

I hunt the tent for mosquitoes and try to kill them before undressing for my bath. They manage to hide until the last article of clothing hits the tent floor—then they zoom into action in great numbers. My baths are speedy affairs.

So is cleanup. Simple meals and simple lodging free me from the enslavements of gracious living. I'm baffled at how leisurely I've felt despite rather close scheduling. Distance from the flurry of town routine must be mind-calming. Though full of things I want to investigate while here, my brain isn't in its customary turmoil. I wonder if there is a way to preserve this after returning home.

Not that this place is without aggravations. But they've produced no stress. They aren't the sort you turn over and over in your mind, that eat at you.

I've mentioned the mice and mosquitoes already—then the other night bigger creatures invaded.

I returned from a sheep drive, worn out, to find a herd of cattle demolishing my campsite. I'd been leaving the gate open because it was so complicated to latch it. During the afternoon about twenty head of cows had wandered in to explore my campsite. Using their heads, they bulldozed my tent and knocked over my makeshift table, then trod on my bread, kicked a hole in the Styrofoam cooler I'd put in a shady spot and scattered my oranges and pears.

I jumped off the horse and ran at the big animals, who were still shoving around my things. I yelled, "Hyahhhh!" and jumped in the air and waved my arms. The cows looked at me with calm brown eyes.

"Git! Go on!" I tried to chase them away but they managed to mill about and meet back between my crumpled tent and the trees. They formed a cluster and turned to face me. I leaped a foot in the air, flailed my arms and screamed, "Boo!" They regarded me with curiosity and did not move.

I glanced back at my borrowed horse and decided she probably had more experience with cows than I did. I ran back and jumped on her.

She headed for the cattle. I turned her to the right to go after a large hereford, whom I suspected was their ringleader, but the mare shook her neck crossly. Several more times I attempted to guide her but she didn't respond well, unusual for her. Once she craned her head around to look at me with what seemed to be disgust.

Finally I decided she knew more about driving cattle than her rider. I let her have her head and off she sped. In five minutes she'd chased the cows far down into the pasture.

Unfortunately, they were back by morning. I awakened to a chorus of moos and what I guessed was a muzzle nudging my precariously re-erected tent. No sooner had I figured out who was nosing my walls, then my temporary home collapsed. A weight of canvas fell against my head. Tundra, who had been sleeping, barked in fright. I crawled out of my sleeping bag and the canvas, blinked into the hazy dawn and muttered at the cows.

I wasn't the only one who was mad. I'd barely pulled on my jacket before I had to face an angry farmer. He'd been pursuing the cows since the previous day and complained that I shouldn't have left my gate open.

Meanwhile, his little boy, about nine years old, picked up a long twig, circled around behind and cows and said, "G'wan now." As a group, the cows sauntered out the gate and up the road.

I spent two mornings on long walks toting tree, wild-flower and bird-identification books. It was pleasant research but I found the books limited help. I discovered trees the book specifically stated grew only east of the Mississippi. Flower identification involved four or five steps and much page flipping. The bird book worked the best if I'd seen the bird so I could find its picture. But frequently I only heard calls I was curious about.

The herders are helpful in that way. I find them to be superior practical botanists and ornithologists. At first I felt I had to check out their answers with the books, but their 100 per cent accuracy has convinced me not to doubt them.

I'd like to crawl into a sheepherder's mind to examine it for clutter. If I walked down its corridors I think I'd find them spacious and empty with room to accommodate incoming images of trees and birds and skies and clouds and flowers and other magical things of God's creation. And then I'd crawl into a town dweller's mind for comparison to see if any nooks and crannies remain for such things. Would a million concerns of modern life leave any open space?

A friend of mine sings this song to her little girl. It comes from the Shakers, members of a nineteenth-century religious community who expressed spiritual feeling with dances and songs.

'Tis a gift to be simple, 'tis a gift to be free.
'Tis a gift to come down where we ought to be.
And when we find ourselves in the place just right,
'Twill be in the valley of love and delight.

About 9:30 or 10:00 P.M. daylight disappears. A friend lent me a lantern so I could stay up and read and write, but I find I'm happy with nature's schedule. I usually have a couple of evening hours to take notes anyway, and by the time the birds and animals quiet down, I'm asleep. Sometimes, owls hoot a lullaby.

Tonight I'm sitting on a rock on the craggy ridge across the road from my tent, watching the day fade. The clouds have orange bellies and, in shadow, the vegetation turns a profound green.

The sun doesn't linger here as it does on the desert or on an ocean. It carefully pulls a pink blanket over the canyon, then falls behind a ridge to switch itself off, and abruptly, the day is over.

It's been useful to be here, not just for information gathering either. I wanted to get at least a brief glimpse of the herder's solitary existence. Among people I know there was little enthusiasm for this trip. When I'd tell people, "It's only for seven to ten days," they'd reply, "You mean a whole week? . . . Well, I suppose if you're working you won't get too bored," they'd concede.

Women, particularly, questioned the idea. One said it sounded ghastly. Men expressed more spirit for it but that may be because they felt they were supposed to like the rugged outdoors. Comments like "Too bad you won't be in a group, or with your family, so you could have fun" and "Be sure to take your radio so you won't go bananas" betrayed their reservations.

One person expressed unqualified delight with the idea. She is a middle-aged woman with nine children. In fairness to the others, I probably should consider that a mother of nine might be more attracted to the notion of a week alone in the woods than the average person.

My husband and I had arranged a calling system because we both would be camping. He was to call my friend DeeDee in Idaho Falls on appointed days and tell her how everything was. Periodically I was to drive to the Bone store twenty-five miles away and call DeeDee to pick up my messages. When I walked into the Bone store the first weekend to ask to use the phone I saw some amused glances. There are no phone lines to Bone. So I've been out of touch.

My garden needs me. I'll be happy to see my colt and Lark. I'll catch up with friends. My house and family will look good to me. How could I be so fickle to let these mountains tempt me away from my happy circumstances?

But I'm sad to think about leaving in another day or two. Maybe it's the beauty. Maybe it's the peace. Maybe it's what a herder told me—"There's magic in these mountains."

XI

Thou Preparest a Table Before Me

"The only religion you have up here is in your heart," Emma Padilla said. "Your own prayers you say that no one knows about."

Formal religious practice, even if once it was important, breaks down in the camps. Churches are far away.

Mexican and Mexican-American herders nearly all have Roman Catholic backgrounds and in many camps Our Lady of Guadalupe and crucifixes decorate walls. Two Idaho priests have a traveling-circuit ministry to sheep camps but the majority of herders say it's been years since they have seen a priest.

The Basques, from a Catholic country, have a pragmatic view of religion and God to begin with. "The Lord will provide . . . for the birds," is a maxim among herders.

Father Ramon Echeverria, Boise, who comes from a sheep-raising family, said pride and self-reliance are strong in Basque character and no mystique surrounds the priesthood or Church authority.

"There is a saying in Basque, 'Every man is a king.' That is the explicit saying. What is implicit in that also is 'Every man is his own Pope,'" Father Echeverria said.

Because of their great sense of personal worth, Basques are very democratic. A card game in the old country might bring together as equals a rich man, a merchant, a farmer and a sheepherder. Priests do not win the respectful title "Father," they are called by their first names just like anyone else.

There is, in fact, a certain wariness toward piety among Basques. When Father Echeverria left for the seminary his grandfather grumbled, "Beware of people who make their living with their mouths—lawyers, politicians and priests."

Echeverria found out about the Basque attitude toward formal religion when he studied and taught in the homeland. Once when attending a mass he noticed a man get up and walk out when the sermon began. The man returned, then exited again when the priest started to consecrate the bread and wine.

Father Echeverria, who did not wear his clerics so he could get better acquainted with people, followed the man out.

"Excuse me, I'm from America and I'm studying religious attitudes here. You don't have to answer me if you don't want to, but I couldn't help but notice your curious worship habits," he said.

"Awwww," the Basque said, "I don't believe any of that stuff."

"Well, why do you attend at all?"

The man leaned over, his eyes twinkled, and his voice dropped.

"Just in case. *Just in case.*"

Eloise Bieter, a Basque-American from Boise, tells the story of a young man who came to America to tend sheep during the great immigration in the early 1900s. His Basque mamma, concerned he would be among a wild, heathen people, made him promise to keep up his prayers.

On the train from New York he looked around at the other passengers. The new land, he wrote home as the train chugged along, appeared to be quite religious. All around him people moved their lips, praying the rosary it seemed. He'd been in America awhile when he learned that the habit he'd noticed on the train had no spiritual motive—passengers were chewing gum.

Because they are universal in outlook, Basques are very ecumenical. A weathered old Basque herder gave Father Echeverria a piece of advice he treasures.

"The important thing is to get the sheep together and the hell with the money," the old man said.

But the number of denominations in America can puzzle.

"Why so many churches here?" Ascencio Burrusco, camp tender for Glen Anderson, Aberdeen, Idaho, asked. "Is confusing. At home, just one."

One old herder Father Echeverria knew had only three years

of schooling but was full of the earth's wisdom. He prefaced his pronouncements with "Every good herder knows," while pointing at his head. When Father Echeverria was starting his priestly vocation the old man put a serious look on his rawhide face, pointed at his head and advised:

"Every good herder knows, you must be smarter than the sheep to herd them."

The Basque provinces resisted at first, then later fervently embraced Christianity. St. Ignatius Loyola, a Basque, founded the Jesuit order. Many Basques return to church habits after they leave herding for in-town jobs.

Idaho Secretary of State Pete Cenerussa, a Basque and former sheep rancher, finds his people "very devout, with a reverence of the highest order." Most Basque festivals, which bring together those still employed as herders and second- and third-generation Basque-Americans of varied occupations, begin with an outdoor mass.

Whatever the nationality or religious background, herders admit their job has some built-in spirituality.

"I don't know why, but you seem closer to Him up here," one man said.

Rube Moses had the unusual experience of having his wife and children with him in the camps, and he believes that experience has cemented his family.

"When you're all alone out there with no telephone or television, ya make do with each other," Moses said. From materials around the camp he manufactured toys—deerhide stretched across a coffee can made a child's drum. His daughters, five in number, raised an orphan coyote. The girls dressed it in doll clothes and carried it around, and when it howled, Moses recalled, it "sounded like ten coyotes outside."

Doris Hunt spent the first year and a half of her marriage in the camps with her husband, John. She remembers that time as one long honeymoon.

"I'd worked on a farm before I got married and always there had been chores waiting for me. To be up there with all that *time* and nothing to do was a real pleasure.

"John would go out in the morning and turn his sheep.

Then we'd follow along behind, hand in hand through the timber. You could take your time, see the clouds and smell the flowers. We had a little tent and sometimes deer would come right up in the clearing. We'd lie quietly and watch them, fawns and their does, only a few feet away. Where else could you get that?"

Solitude brings some herders to a poignant feeling of "just Him 'n me."

"It clears your head up here," one said. "Problems don't go away, they just don't seem so big."

They appreciate it when someone understands those emotions. Bill, a former camp tender, remembers one late evening at a herder's camp. Bill usually wasn't so late, and Pete, the herder, already was in bed. The camp tender woke him anyway—he figured Pete would want his groceries and, besides, he'd brought a letter from Pete's boy.

Pete's grown son lived in California and had a good job in a lumber mill there. He wrote his dad often and Pete lived for the letters. Pete couldn't read or write so Bill always ripped open the envelope and read the letter, repeating passages on request.

That night there was no coffee for Bill because the fire had died, but Pete was eager to hear the letter.

Bill read through it, the news of grandkids, vacations, car repairs and weather. At letter's end he read, "And God bless you and be with you out there."

"Would ya read that last part again, Bill?"

"He says, 'And I pray God will bless you and take care of you out there.' "

The old herder nodded. He pulled the covers up over his shoulder and blew out the kerosene lamp.

"Thanks a million for that, Bill."

So many changes.

The rancher leaned against a sheephook, gazed up the mountain and thought about that.

He hated the power lines which stood beside his formerly isolated sheep trail. The lines carried electricity to summer cabins and condominiums tucked in the woods. He easily could have tracked those summer visitors; they left signs everywhere —beer cans, Kleenex and candy wrappers.

At one time an unwritten code of brotherhood among western herders made it standard policy to leave sheep camps unlocked. Packaged soup, macaroni and canned goods in the cupboards were for the taking by hungry visitors.

But that was when the land was quiet, before the rev of motorcycles engines and scream of chain saws. Now everyone locked everything, and still tires, rifles, radios, clothing, bedding, screen doors, even doorknobs, disappeared from the camps.

They took sheep too. A few knew how to hurriedly kill and dress an animal and escape before the herder noticed. Some took live lambs home.

But most intruders just chased for sport, swerving motorcycles and four-wheel-drive vehicles after panicked sheep until the animals dropped from exhaustion. The rancher remembered that the previous year he'd had to finish off lambs with broken legs, cuts or internal injuries. Some nursing ewes died in their tracks from terror.

One rancher had told him he feared serious trouble reminiscent of the range wars if angry herders ever started to retaliate. Several times, the rancher recalled, he'd scared himself—he'd had invaders in his rifle's sights, but he backed off. He knew many usually peaceable herders who swore hatefully against the "two-legged coyotes."

Motorcyclists, it seemed to him, could not grasp that in fenced pastures gates should be reclosed. Owners of four-wheel-drive vehicles seemed gaily unaware that side paths cut by their machines caused erosion. On top of the real damage, their paths caused herders one more aggravation. Animals tend to follow such paths and it took more herding to keep sheep going right.

Sometimes it seemed like sheep had human enemies on all sides. Not all enemies came on loud machines—some came afoot, freeze-dried food in their backpacks, and some came in quiet gray cars, their agency's name lettered on the doors. The rancher blamed environmentalists, along with bureaucrats, for some hurtful decisions.

Pressure from protectionists stacked the odds against the sheepmen in their battle against coyotes. Before 10–80, a tasteless, odorless poison, was banned, sheepmen were gaining on the coyotes. Ranchers swear that the poison caused no verified case of harm to humans and that trapping, shooting and other legal methods are insufficient to keep up with the burgeoning coyote population. Environmentalists feared the poison indiscriminately harmed eagles, marlin and the scavengers that fed on dead coyotes. But the rancher had heard even forestry officials admit that the ban was too tight, and with proper precautions the poison could greatly help sheepmen and do minimum ecological damage.

On National Forest land grazing allotments had been cut back, often 50 to 100 per cent. As the rancher saw it, meadows which previously sustained thirty thousand head now got only a token nibbling by five or ten thousand head and by August unused feed stood knee high. The Forest Service defended the policy by pointing out it had salvaged badly overgrazed ranges. Still, the rancher disliked blanket policies.

Along with increased hay costs, grazing fees had spiraled. They were five or six times as high on federal land as they were fifteen years ago. State land fees were ten times as high and private leases kept pace.

And the regulations—pages and pages and pages, the gist of which he had to convey to herders who often didn't speak English.

It made the sheepman smile to think that sheepmen might yet get the last laugh. On a forest in Washington State, forestry officials had excluded all sheep from an area. Vegetation got high, fires erupted and conditions became unhealthy, and government officials sought out sheep to come back. Rumors said that other places which had banned sheep might do the same. Sheepmen knew that if properly monitored, sheep left land in better shape by pruning and fertilizing.

Other encroachment brought its own headaches. Towns spilled over into formerly secluded areas where the rancher wintered his sheep. Plenty of desert was left, but trailing got progressively more complicated. Trucking, the expensive alternative, loomed in the future. The irony, he thought, was that people left city neighborhoods to absorb the jobs of rural living, then became incensed if a cow or sheep appeared nearby.

One by one, the rancher had watched the big sheep outfits fold. He could remember when it wasn't unusual to run ten or twenty bands. Once, there was gold in that fleece. Now five bands was a big outfit.

Before the Taylor Grazing Act of 1934 and before homesteaders claimed western land, fortunes were within reach for the diligent. A herder one year could be the owner-rancher ten years later.

In the early 1900s. Frenchmen ran large bands through the Idaho hills. They'd start trailing in Oregon, moving bands of wethers five or six thousand strong. They would graze through Idaho en route to Kansas City markets—the trip took about three years. Newcomer sheepmen complained they could scarcely find a range to lamb on if Frenchmen were coming through—they occupied every hill for miles.

Itinerant sheepmen made big profits. Even men who ran bands on leased or purchased land did well because feed and labor costs were small.

In early days it was popular among herders who aspired to bigger things to take their wages in sheep. Then their animals would run along with the boss's. Ranchers liked the arrangement because it gave herders incentive to take good care of the sheep. Before the law changed, good feed belonged to whoever got there first. That transient approach did not foster affection for the land, and ranges were abused.

Whenever sheepmen in Idaho discussed rags to riches stories, Andy Little's name came up. As a young man he spent his small resources on passage from Scotland to America. When he got West he hired on herding sheep. He acquired a few head of his own, then more and more and more. His operation at Emmett, Idaho, continued to expand and Little became the largest individual sheep owner in the United States. He ran twenty-two bands, and according to the legend, the banks didn't have "a penny agin' 'im."

The rancher had known personally an immigrant herder who made good, and his story was typical of many early sheepmen.

For ten years he lived in the hills with only the sheep and a dream. He had no costly habits like drinking which might have depleted his savings, and he seldom even took a trip to town. He squirreled away every cent he could.

At the end of that time he bought a band of his own. With good management he got another.

But a series of reversals sent his fortunes another way. He lost everything.

He sought out his former employer, got back his old job and returned to the hills. Again he lived frugally and banked his earnings for ten years until he was able to buy sheep again. He prospered and until his death was the respected owner of several bands and a big ranch.

Idahoans identify their substantial Basque population with sheepherders, but it was rather accidental that Basques cast their lots with sheep and Idaho. Boise has the largest Basque settlement in the United States and Basque festivals and complicated surnames are a fond part of Idaho culture. But originally California and gold lured the Basque immigrants.

News of riches in the new land reached "Euzkadi" (which the people call their homeland) at a time of economic and political hardship. In the Spanish provinces, where most Idaho Basques come from, Spanish reign had robbed the people of a long independent and democratic tradition. The Basque language, which experts believe to be the only extant prehistoric tongue, was suppressed in the schools. Mandatory conscription into the Spanish Army gave young men more motive to leave. And on the tidy family farms, a system remained intact that passed entire holdings to one deserving son or daughter. Other siblings had to seek a living elsewhere, and the new land offered possibilities.

Rumors of easy fortunes in California were exaggerated, and men had to take other work to feed themselves. The population in boom towns had to eat too, mutton prices soared, the sheep business prospered and many Basques became herders.

A few French Basques had experience herding sheep in the Pyrenees, but they had tended small farm flocks under conditions much different from those in the American West where huge bands ran on immense empty spaces. Most Spanish Basques had no experience with sheep.

So experience with sheep was not why ranchers liked Basque herders. It was their cultural commitment to hard work, their legendary honesty and physical strength. The rancher could still picture the Basque he once employed who would pick up and carry on his shoulders a weary ewe. Ewes weigh about 150 pounds.

The seven provinces of the homeland (four in Spain, three in France) sent new herders to replace those who left for other jobs or became sheep owners themselves. Newly prosperous Basques honored strong racial and family ties and sent home for brothers, nephews, cousins and neighbors to help in their outfits.

Immigration policy changed in the 1920s, and good Basque help wasn't as available to ranchers. In 1934 the Taylor Grazing Act regulated range, and handy fortunes moved out of reach.

Years later wool growers pressured Congress to make accommodations for sheepmen, and Congress passed an act that permitted alien herders to immigrate under term contracts to livestockmen. Ranchers still bring over Basque help, but a vastly improved economy in the provinces means fewer Basques want to come.

Some Basques who herd still cling to thrifty ways, and when contracts expire they have enough money to go home and buy businesses. Some herders send wages home to help families.

Some find the adopted land becomes home and go to the provinces only for visits. They seem reluctant to admit they aren't going home—they stay in the hills until it's too late. A rancher knew a herder who still hoarded his money (it was rumored he had $100,000 in the bank) and he was in his eighties.

Good men weren't coming along to fill the boots of the dependable Basques and the veteran herders. It seemed to the rancher that labor problems had worsened during the last ten years. A good man could affect profits so much—a bad one did the same.

Within his own outfit methods had changed too. Until recent years the rancher never had staked a horse. At one time he'd hobbled the ones that wandered, but heavy-duty hobbles were hard to find in stores anymore. Many sheep-camp horses were good about staying around, but untethered horses weren't as safe as they used to be.

At one time sheep-dipping was a regular event. Companies with huge portable vats and portable corrals visited the sheep outfits to treat sheep against ticks and lice. They'd swim sheep through an insecticide, completely immersing them at one point, and it was quite successful in keeping pests off sheep. This was particularly important for outfits that ran sheep on the desert.

Such companies were now scarce—too few sheep outfits to cater to. Many outfits sprayed their sheep, which was less effective because it was hard to soak the wool. The government had banned dieldrin, a chemical effective against parasites, and another product, Kreso, had been weakened.

Even equipment had changed. The sheephook he leaned against was a modernized counterpart of the ancient shepherd's staff. It was longer, with a fifteen-foot handle and a metal hook on the end. It was designed to hook sheep by the leg, rather than the neck.

Even that tool wasn't used much anymore. And not every wagon had one hooked to its side like in the old days. Some men preferred to rope animals and some used electric prods. A few old-timers still used canes to help load sheep, but that was unusual too.

The rancher peered up the mountain, then started to walk. Even after a lifetime in the hills, mountains could still trick him. It either was an optical trick or persistent optimism which made it seem that the mountain rim was just above the next switchback. Even knowing forest and rock landmarks as he did, he sometimes was surprised to find himself facing another climb when he'd thought the last one would put him on top.

The grandeur and challenge of the mountains hadn't changed. And that was a comfort.

XII

Thou Anointest My Head with Oil

The dog is the sheep camp's barking, tail-wagging greeter. He rushes out and jumps on me, rests his paws on my waist and coaxes me to rub his head.

He has an unusual situation—he is both above and below the conventions of town dogs. At a dog obedience class his exuberance might be considered unmannerly. His matted coat and burr-tangled tail certainly wouldn't pass muster with a dog-show judge. He doesn't know a single trick. He gnaws with relish on animal entrails polite city dogs would refuse.

At the same time he enjoys a status his city cousins couldn't hope to attain. He is disciplined by his own loyalty and desire to serve rather than by popular training methods. He insinuates himself into his master's life, often becoming the herder's chief love.

Once again, I find that camp life's simplicity figures in the picture. Here, dogs' claws do not threaten fine clothes and fragile stockings—disagreeable snacks won't be regurgitated later on a sculptured shag rug.

But it's more than that.

I heard a story about a hard-working Mexican immigrant and his dog. For twenty years the Mexican worked for one outfit, distinguishing himself for his reliability and physical strength. Though the herder had no bad habits to drain his savings, he seemed to accumulate no material possessions and he dressed in worn clothes and shabby footwear. Finally his boss learned that the Mexican was sending his brother through school—first college, then graduate school.

Failing health finally forced the herder to leave his job. He sent for his brother in California.

The brother, now a college professor, arrived at the sheep camp in a gleaming car of prestigious make. The rancher was there to see the herder off and express his gratitude. He also pointed out that the dog the herder was clutching belonged to the ranch and was important to upcoming work.

The short, frail herder shook his head vehemently and announced he was taking the dog, and that was that.

"What else do I have?" he asked. "After all these years, what else?"

Cradling the dog and glaring at the rancher, he backed to his brother's car and eased himself in. The herder's brother looked distressed as the muddy, drooling mongrel danced happily around on his velvet upholstery. He and the rancher exchanged a helpless look, then the brother drove off.

The Idaho Falls paper carried an account of a fatal shooting at a sheep camp, the result of a fight over a dog. A herder left his dog behind when he quit an outfit. Later he returned to get it but the herder who replaced him said he needed the dog and claimed it belonged to the outfit. One man pulled a knife, the other a gun. Seconds later a bullet tore into the herder's chest. Sentimental dog lovers may have smiled at the story without understanding the raging emotions and desperation behind it.

I've been thinking, lately, that in our technological culture we no longer grasp the deep attachment with beasts that comes from dependency on them. The Indian in Mexico who brings his harvest to market on his burro, the Egyptian whose water buffalo turns a wooden lift to raise water from the river and the herder of the American West may share a relationship with animals lost to most of us. Animals can be our friends, our playmates, sometimes our work mates, but seldom are they our partners in survival.

As a horsewoman whose main love is trail riding, I thought I knew how much a person can treasure a good animal. Last summer when I lost my beautiful buckskin mare, Buffy, my world collapsed. My best friend drove out to spend the day with me because she sensed how desolate I felt. I grieved for Buffy not only because of the long relationship and special communication between us, the way we "clicked," but also for practical reasons—she was more cautious, alert and sensible than horses I rode after her death. I understood how important a good horse is to enjoyment and safety.

But new insights have come to me during my trips up here, and they crystallized during a night ride through the mountains.

When I camped near Brockman I had the use of Dolly, the late Joe Carter's horse. Her owner took her out of retirement to lend to me because my own mare was too green and young for mountain work. Dolly wasn't used much this spring, but she runs with a band in the mountains and stays in good condition.

I liked her when I first saw her. Old horsemen say a horse's face tells its nature and intelligence—they study the location of the eyes and whorl of hair, shape of the nose and expression of eyes and ears. Dolly's eyes were kind, ears alert.

Gas stations were far away so I resolved to drive as little as possible during my stay. It would be a waste of precious working time to make trips to the valley for gas. So Dolly carried me and my water jug to a nearby spring every day, to the Brockman Guard Station for dawn appointments and on sheep drives. In depending on Dolly to help me do my business, I had a glimmer of preautomotive days. And every day at dusk she carried me beside the creek for a ride I didn't even pretend was business.

One afternoon I'd laid my poncho, Dolly's oats, extra jeans and my toothbrush beside a log—I was preparing to ride up the mountain to join a pack trip. Sand- and silt-covered underwear I'd washed in a nearby hot spring hung on a limb to dry. As I knelt in the grass sorting things I heard the gate creak with someone's weight. I turned to see a man crawling over it.

Despite my isolated location, I'd had a visitor every day. Camp tenders, cattlemen and sheepmen had dropped by to say hello and satisfy their curiosity about who was visiting their territory. One jolly trucker stopped to ask directions, and stayed for lunch. Though I was a bit coy about giving information, I think it was obvious that I was camped alone. So far I hadn't felt the least apprehensive.

But I froze with fear as this man, thin and about thirty, walked toward me. He was out of the mold of others I'd met that week. Though dressed in soiled work clothes, he didn't have a look of honest labor. Neglected patches of hair darkened his jowls and upper lip, and his eyelids drooped at the outer corners. He stared vacantly.

"I saw you down at the creek doin' your laundry—ya camped here 'er somethin'?" He eyed the flap of my big tent and asked, "You alone?"

Maybe friendly Tundra would sense my distrust and for once behave fiercely. I strained to tell her, with telepathy, to drop to her stomach and growl. She bounded over to the man, wriggled with glee and sat down on his foot.

He mumbled a murky story about being lost and he thought I could help him. His eyes shifted from the gate to the tent to the thin road, which curved away into an empty distance.

I offered my own lame story about leaving shortly on a trip with a crowd of people, some of whom would arrive momentarily. Instinctively I moved over to untie Dolly and lead her away from the trees.

"Which way your friends comin' from?"

"Down from Brockman."

"In a truck?"

"Yeah."

I wanted to slip the lead rope over Dolly's neck to make a makeshift rein, but the man was close to her on the other side.

"I was jus' up to Brockman and there ain't no trucks up there. It's deserted."

It was bad news to me that he'd already cased the area. Brockman is merely a Forest Service Guard Station, usually abandoned, and I regretted my story.

"Well, Dad's coming in a truck, but my brothers are coming on horseback." I tried to gurgle with enthusiasm, but my voice cracked. I squinted into the trees behind the tent as though I had just spotted my mounted brothers. Always I'd wanted a brother, but never more than at that moment.

If only Dolly were saddled and bridled, I thought, I would feel calm. I'd coolly swing into the saddle and wait for the man to leave. If the situation became uneasy, Dolly and I would spin around and be off through the trees.

The enormous solitude I'd grown so fond of now was my enemy. I could waste no thoughts on someone happening by. Quietly I put my elbows over Dolly's back and pulled myself up into a sitting position. She took a step and I whispered, "Whoa!" She whoaed.

The man, arranging dirt with his scuffed toe, looked up, puzzled. I feared it seemed odd for me to be boarding an unbitted, unsaddled horse. I flung a booted foot over her neck, rested an arm on her rump and tried to look nonchalant.

He made no move to leave. I chattered nervously, inviting him to stay to talk to my brothers, who knew the area and could help him find his way out.

Meanwhile, I plotted that if I had to, I would give Dolly a mean kick, and hang on. She wouldn't need more than that to take off at a run. I worried she would whirl and run down the hill behind the tent, through the swamp, jump the creek and head toward her home range. With nothing in her mouth to slow her, I suspected she'd fairly fly. The meadow had a cow fence along the road, but behind the tent was a vast, unfenced land. The problem would be staying on until I was at least a safe distance. Yet, the choice seemed clear. If things looked chancy, I'd take my chances on Dolly.

Why was he so silent? He tapped his feet and wriggled his face nervously. He seemed a bit persuaded that horsemen soon might appear because he watched the trees expectantly. Finally he muttered that he'd drive around a little to see if he could find his way.

He started toward his beat-up truck parked on the road. I tossed my saddle on Dolly and slipped a bit into her mouth. The moment my belly cinch was secure my alarm evaporated. I had miles to ride alone on the road, but if the disreputable-looking man reappeared, Dolly and I could detour where no truck could follow.

Hastily I packed my things. Doing my laundry and hair-washing had been a time-wasting vanity. The men, sheepman Roy Cooper and herder Ab Waters, had warned me to allow enough time to travel the eighteen uphill miles. The eye-level sun reproached me. The stranger in the blue truck had closed my option of leaving early next day and catching my party, who would be slowed by sheep and pack horses.

Wrestling the gate shut and wedging its makeshift latch into place cost me several minutes. Then Dolly took only a few steps and socks and soap slipped from my poncho. I dismounted, stuffed them in, jumped on and we took off at a run.

A knife clinked on the rocks and toothpaste flew into the weeds. Again I collected things, but the scene was repeated after we'd cantered a few lengths more. It occurred to me as I searched the ground for spilled objects that horsemen making fast exits in Westerns never bungled it like this.

An oat sack poked over my saddle horn carried my camera, lenses and film. But the pack behind the seat was too wide and unwieldy. I repacked my bundle and dumped Dolly's oats on the ground. She turned to me with what I perceived as a dirty look.

This time we cantered away without losing anything. By now the sun already nudged the top of a distant range, and a pink glow fell on the valley. As we ran past a pool, a mamma duck beat her wings against her brood to hide them behind weeds. On a ridge, two fawns stepped between the rocks on a path to the creek.

This was my favorite hour of the day in the mountains, a time to guzzle the air. Twilight fragrances are prettiest, and the little valley trapped and concentrated them.

The first seven miles were a gentle climb, so I made the most of them and let Dolly run as much as she wanted. She felt frisky in the cooling temperature.

After we passed the guard station we turned up the mountain. I was dismayed to see that only a half circle of gold remained on the horizon, and we had ten or more uphill miles to go.

Dolly wanted to trot but I restrained her—I didn't know if she was in condition for it—and besides I thought we might need to reserve a push for later.

We followed the vertical road that winds up to Skyline Ridge. The sun was gone but an orange-brown aura clung to the hills. One pleasant thing about the steep switchbacks was that each time I thought daytime was over we conquered another ridge and gained a slight reprieve—a bit more light. The view grew more spectacular—ranges in purple silhouette lay beneath us. The quiet was soothing. But my spirits sunk each time we finished a climb and I saw how much mountain remained.

At last atop the Skyline Ridge the road leveled and Dolly wanted to move out. She broke into her odd gait, a running walk, and hope filled me again—the trees were speeding by.

From this elevation I watched the last of daylight melt away. We had about eight miles to go. The ground was a black sea but an outline of tree crowns marked the road. Coyotes began to sing.

We'd hurried along for a long way when I saw a dome-shaped silhouette. I squealed and thumped Dolly's neck—we were there.

Dolly broke into a gallop. A light went on in the camp and a man's head appeared in the doorway.

"Hi!" I yelled. "Boy, am I glad to see you. I'm sorry I woke you." From a distance I rattled off an explanation of why I was late. I was so relieved to be there I couldn't stop grinning.

I looked down at the man. It wasn't Roy or Ab. It took me only a minute to figure out he was a young wetback I'd heard was in the area. I held my forehead and groaned. This meant I still was about six miles from Roy and Ab's camp on Commissary Ridge. No use trying to explain my error, the boy couldn't speak English. I whirled Dolly around and we thundered out of the clearing as fast as we'd arrived. When I glanced back I saw the man bent sideways around the camp's wall, staring after us. Ab and Roy told me the next day I might have cost the man his sanity—lonely herders sit at night and dream that female visitors somehow will materialize. Ab thought the poor man probably thought he'd had a vision.

Now the night was pitch black. Tree outlines were obscured and I had no idea if we were on a road or if we were winding through trees. All evening I'd resisted Dolly's urges to take crosscountry shortcuts. I knew she was well acquainted with the area, but I wasn't sure if she understood our exact destination, or if she might head for some other familiar place. Now I had no choice but to let Dolly choose the way.

My reliance on her went beyond her knowledge of the route. Though I worried about the late hour because I'd have to pitch my tent in the dark and I'd probably awaken the men, I otherwise was enjoying my night ride. On a foolish horse the situation could have been unnerving. But movements in the trees, night sounds and an animal dashing in our path were things Dolly handled with poise.

"I'll take Dolly for the night run, Bart—the mail must go through."

"Yes sir, General, Dolly and I will go for reinforcements."

"Take Dolly and ride for the doctor, Jeb. Ma's in a bad way."

I had a glimpse of how it used to be. I felt deeply grateful to Dolly and my ride was on a cool but mild night. How many had relied on good horses during true survival situations! Then, to be tossed or have a careless horse fall would mean pain or death.

Maybe our sophistication cheats us of a valuable alliance with God's other creatures. Herders still have it, with their dogs and horses.

David composed his verse in times when man's dependence on animals was personal and dramatic. That may be significant when we remember that David drew parallels between the man-animal association and the God-man relationship. To see ourselves as passive sheep to our God-herder may underrate our role in the partnership.

I expected Dolly to tire, but she kept up her ambitious pace. Tundra pitter-pattered at our side, sometimes swishing off through nearby bushes. Finally, when I heard dogs start to bark in the distance, I gave in to some weariness myself.

Dolly managed a burst of speed for the last stretch. The camps were darkened, but as Dolly's hoofs thudded near them a sleepy voice moaned, "Who the hell is that?"

Ab and Roy, pulling on shirts, came to the door. I apologized. Ab petted Dolly and led her away. Roy pointed me to an empty sheep camp so I wouldn't have to pitch my tent, poured some water into a bucket and lighted the kerosene lamp for me.

The camp, especially the bed, looked wonderful to me. I unrolled my sleeping bag, skipped washing and flopped into bed. I was nearly asleep when I remembered I'd dumped Dolly's oats, and I had no reward for her.

XIII

My Cup Runneth Over

At twilight, shadows emphasize details in the sagebrush and deepen the green on the hills.

Sheep mill, preparing to bed beneath a house-size lava rock. Low sun lights the black-speckled rock in orange. Distant ranges turn magenta, trees become silhouettes.

Firs and cedars and maple bushes exhale on the evening breeze. Mountain orchids, side by side in an acre field, send out a delicate scent. Catnip sprinkles the air with fragrance of mint.

Sitting on a rock ledge, the herder listens. A woodpecker thumps a nearby tree. Partridges call in rolling voices and sandhill cranes sound their creaking cry. Nearby, the horse munch, munches a clump of wheat grass. Sheep bells tinkle.

Then a coyote moans. In a moment, an answering howl pierces the serene dusk. The herder gets up and walks off to get his gun.

He hates them. They take his best, fattest lambs. He spits out curses when he finds their telltale bites on dead animals. He rages when they come back to kill more while dead ones go uneaten. He dreads their growing numbers.

They kill by clamping their teeth into the sheep's neck and

holding on. Sheep die, usually of suffocation, in about fifteen minutes. Ewes sometimes defend themselves and their lambs from attack. But a ewe might repel one coyote several times and then be killed later by others.

The herder dislikes any predator. Cougars sometimes take lambs. The cats leap onto their prey, leaving bite marks on their victims' heads. Yet, the herder admits big cats fascinate him.

Eagles can be a problem, particularly to outfits who lamb on the range. Talon marks on carcasses betray the killer. But the herder can't help being impressed by the tall, golden eagles. Once he thought he was watching a short man on the horizon, but when he looked through his binoculars he saw a strutting, three-foot-tall eagle.

Magpies kill under certain conditions. If sheep stay in one location a long time they shade up for longer periods and give magpies a chance to peck holes in their backs. Flies lay eggs in the wound, and when maggots hatch they go into the body. Crows, too, can cause losses—they peck out eyes of young lambs.

Bears slash a path of destruction through the flock. Particularly when sheep shade up in the trees they are vulnerable to bears; sheep on the outside of the band don't know they can escape into open country, and only bunch tighter. But bears are a seasonal problem, at least.

Coyotes are a year-around menace. In midsummer when they are eating chiseler squirrels they ease up on lambs a bit. But squirrel and rabbit populations fluctuate from year to year, and anyway, coyotes sometimes kill without regard to appetite. This year a herder lost fifteen lambs in one location, and not one was touched for food.

Last year the herder's boss had a $4,000 loss to predators, and the coyote was the biggest killer. Almost a million adult sheep and lambs in the West were lost to coyotes that year. This spring the herder lost two hundred lambs out of two thousand from lambing time to shearing. Summer losses were running equally high.

Coyotes hate men as much as herders hate them. They so dislike, or respect, human scent that some herders use unlaundered garments as deterrents. Left hanging near the sheep, clothing strong with human smell will keep coyotes away.

The herder doesn't know how many coyotes out there tonight threaten his sheep. Two can sound like many. He once peered down the throat of a trapped coyote and saw its tongue flip-flop as it howled, and he better understood how one can sound like five.

The moon is climbing but daylight still peeks over the hills. The herder cradles his gun in his arms and patrols around his band. The coyotes are quiet now.

Stinging nettle and lavender-red fireweed line a path to the stream. The herder hikes down a rocky northern slope controlled by Douglas fir. When he leans over for a drink, white flowers the size of a man's thumbnail, growing in a quiet pool, come into view. He hears Judy bark.

He bounds up the hill. Through the trees he sees Judy, barking and snapping at a coyote. His gun is against a tree back at the stream.

He grabs a log and runs in at the coyote. The coyote whirls to face him, snarling. Judy runs in to worry the coyote from behind.

The coyote winces when the log bangs against its ear. Its eyes flame and saliva drips as it snarls. It shows its fangs, then crouches to spring. In midair, the log smashes its chest.

Judy rushes in to bite and distract the coyote. The herder clubs it again and again. His muscles tighten after each blow and he hisses through his teeth.

The coyote thrashes and threatens. It wants to flee now, but Judy and the herder have it cornered. A few hard blows more and the coyote falls over on its side. Blood runs out its mouth.

Sweat runs off the herder's face. He reaches for Judy's head. He wants never to kill a coyote this way again.

And yet, it was satisfying. His boot kicks at the dead enemy.

John set his scuffed boot on the camp's weather-stained wood stair and stepped out into the morning. He balanced one hand on the metal door latch and kneaded his forehead with the other.

His brain swelled against his skull and his stomach rolled and bucked. Something else internal panged too, and in a vague way he knew it was his heart.

He'd had two days off in town. He'd gone in for his birthday. The boss took him over the washboard roads to the valley and dropped him off.

For several weeks John had nurtured a desire to go to town and be with other people. Finally he told the boss and the boss brought up a relief herder. John asked for a couple months' wages.

The boss gave him the money in cash because herders have trouble cashing checks in town and said "Be careful" when he let him out on a downtown street.

The town was a clean, family-life place with a friendly business community. The friendliness did not extend to work-worn trousers, shiny skin and faltering English. While the herder tried out new binoculars at a sporting-goods store he felt the stylish clerk's impatience. Defensively he pulled out several twenty-dollar bills and spread them on the counter before he'd had a chance to look over all the models.

He walked out into the Friday-evening bustle. While waiting to cross the street he noticed two giggling, scampering children beside him. He smiled down at their pink round faces. So hungry he was for the sounds of childish voices he squatted down and took the little girl's hand. When he turned his face to the children's mother he saw her uneasiness. He stood up and joined the crowd in the crosswalk.

On a street where businesses write their names in blinking light bulbs, the herder pushed through a clouded glass door and into a dark room. Murmured conversation came from the counter and a twangy lament played on the jukebox.

The bartender greeted the herder with a wide smile and simultaneously a waitress was beside him. She touched his forearm. The hurt of nights alone collected in his stomach and throat and he looked at her with hope. His father, long ago in another land, had told him if you want to hate someone it's all right to do it on the street, but if you want to love someone you must slip away in secret. He could use some love now, either in secret or before the crowd's gaze.

Dance? She pranced on tiptoe and held her arms in a circle to show what she meant. He nodded. Drinks for the house? Her bare arm gestured around the room to the fifteen or twenty people on stools or at tables. He nodded again and pulled out a hundred-dollar bill. She smiled and crinkled her nose and took the money. When she returned a long time later he wondered where his change was, but she pulled him to his feet and fastened his arm around her waist.

Above the woman's sweet-scented hair he could see Francisco across the room. Francisco was a handsome young herder he met in here last year. He had worked a long time for one rancher and he owned his own trailer, parked on his boss's place. In the winter Francisco assisted with maintenance around the ranch and sometimes helped with farm work in spring.

He had bragged to John of his thrift. He had been able to accumulate many things to furnish his trailer—records, a stereo, color television, big books with color photos, dishes, a watch, clothes and a sheepskin jacket.

This spring he'd heard of Francisco's misfortune. Francisco's outfit had hired two Mexican brothers. One had a car, a station wagon. The other had a pretty wife, shy, small and young.

The first night in their new job the brothers invited Francisco to their camp for coffee. The brothers told Francisco they would all be friends—they would plan activities together for summer. They all laughed and joked in their own language and when a long silence came the younger brother invited Francisco to be alone with his wife. Francisco looked at the beautiful girl with surprise. He jumped at the chance.

When Francisco returned to his trailer some time later it was picked clean. The brothers had crammed full their station wagon with his stereo, his coat, his $140 boots, books, dishes and watch. His money was gone too. Francisco's boss called the sheriff but the trio never were found.

Which made John think about Louis. Louis was an old Basque herder he met several years ago. One time in the hills Louis found an old wallet with two twenties in it. John kept teasing him, "When are we going to Soda Springs to spend that money?" But Louis insisted he was going to return the money to its owner.

At year's end Louis had $1,700 coming. He laundered an old sugar sack and asked his boss to give him his wages in twenties. With his sack of bills, he went to town.

The next day his boss got a call. The Idaho Falls police had

Louis in jail for drunkenness. They found a wallet on him with a name card in it and the sack of twenties and thought he'd stolen the money.

The boss went to the police station and verified that the money belonged to Louis. While there he and Louis called the wallet's owner.

The owner drove down from Roberts, a small town twenty minutes north. He looked at his wallet, at the unshaven herder and left without a word of thanks.

Herders didn't fare well in town, John knew. So when the woman drew his arm tighter he worried that the happiness he felt might be like the beautiful flower of the bitterroot, which blooms as the snow melts, but quickly withers.

Back at his camp, John rubbed his head. He couldn't remember much about the rest of the evening. The more he drank the more drinks he supplied for others. He might even have gone through his money and run up a bar bill, he didn't know, he'd done it before.

He'd left with the woman. He hoped he remembered wrong that it had cost him $100 to go with her to her place. About 2 or 3 A.M., when he was hurting from too much to drink, she shook him and told him if he wanted to stay through the night it would be $100 more.

John called his dog Molly and started the sheep down to water. He worked his way through the chaparral following Molly with deliberate steps. When he bent under an elm branch his stomach pitched, and when he straightened back up pain stabbed his head.

How often herders were fools! Old T.K., in his early nineties, was in a bar one night with a girl who looked underage. She'd made off with the shiny big-engined car he'd saved so long to buy and T.K. could only shrug when people asked about it.

T.K.'s boss, worried about what would happen to him when he wasn't able to work anymore, had started a savings account in his name and kept the passbook himself. But T.K., desperate for drink one day, managed to talk the bank out of the money and drowned his savings in four days.

Was today Monday? Or Sunday, or Tuesday? He shrugged.

A yellow mountain canary landed, bobbed on a thin branch and began to warble. In nearby weeds John heard the rustle of chiseler squirrels and when he looked over he saw two fat-bellied ones playing tag, their short-rope tails held in a curve.

His sheep lined up along the stream, fleecy sides ballooned from early-morning grazing. They, at least, had been well cared for in his absence. He remembered a herder in his outfit who once deserted his sheep and hitched a ride into town with some picnickers. When the boss found him at a downtown saloon he asked, "Who's with the sheep?"

The herder, happy from whiskey, told him, "Oh, I left 'em with the Lord. I figgered he'd do a better job 'n me anyhow."

John looked off to the east at the Teton's jagged blue outline striped with white unmelted snow. He glanced south at the Lost River Range's faint shadow. Clumps of aspen embroidered the near hills with irregular dark-green patterns.

He sat down on a rock while the sheep started a slow munching pace back up the hill. He pulled his arms around his knees and rested his cheek on his Levis. He marveled that in the middle of such total pain a little happiness started to trickle up to his heart.

Overhead a chicken hawk screamed. On a gentle thermal it soared, hunting. Its wing tips pointed slightly up, and orange hues on its underside made a ribbon of color as it circled.

In a moment it dove, plunging to weeds about fifty feet away. John heard a feeble screechscreechscreech from the hawk's victim, a chiseler squirrel. Other chiselers would be tearing off for holes and for a few minutes the woods would be quiet. The squirrel's brown body wriggled against the locked talons as the hawk flew off into the morning.

John's body felt enough better that he knew he'd live. Later on he would make promises to himself. In all likelihood he'd break them. Sometime he'd again find himself on a risky, exposed slope. But the morning was too pretty to think about it.

XIV

Surely Goodness and Mercy Shall Follow Me

The grass went first. It pulled a brown shawl around its blades and tried to hide from the sucking heat. Flowers sagged, their petals gone. Wild strawberries struggled to bear fruit, but when the berries finally appeared they were tiny and juiceless. Chaparral on the mountain's rim did not turn green under the spring sun but waved purple leaves in distress.

Scanty spring rains had followed record low snowfalls and the land whispered the hateful word—*drought.*

The herder touched the brittle grass and wondered how much longer it could feed his animals. Its protein content lowered, the ailing grass did not satisfy the hungry ewes and would not fatten young lambs. Already the animals had started to feed on browse and their ravenous nibbling threatened to damage shrubs for next year.

Springs that always had gurgled with melting snowpack only trickled. Wide creeks that normally roared through canyons lapped along quietly. For the first time in anyone's memory, Brockman Creek was dry. The herder worried what August would bring. To keep the sheep on dry feed and haul water to them would break his outfit.

On the radio, officials proposed emergency measures. They would seed clouds. But no clouds formed. Congressmen would seek federal money to compensate people in agriculture for their losses. But the herder knew that if hard-pressed sheep outfits went under, they never would recover.

Indians promised to coax the Great Spirit with rain dances. In the valley, congregations knelt in the pews while ministers entreated the Lord. A Boise priest initiated a prayer chain. Signs everywhere, on drive-in restaurants, insurance companies and bumper stickers read, "Pray for Rain."

In the farmers' dry-land wheat fields, stalks had stingy, miniature heads. Willows beside ditch banks wore faded crowns. The long, probing roots of elderberry, snowberry and bitterbrush managed to find moisture, but what if the drought continued—what about next year?

Cheat grass in the lower pastures already had dried up, turned purple and blown away. When sheep ran out to familiar meadows they searched and pawed and later moved on with empty stomachs and dust-covered muzzles. Slender wheat grass and mountain brome milked small drinks from the soil and fought to live, but the herder knew its days were numbered.

At higher elevations, meadows bloomed a month early because there was no snow cover. Vivid yellow flowers of dock, a favorite sheep food, painted the meadows. In ten days they would be gone, before the sheep arrived.

When a patch of gray shaded the horizon, the herder stood up, looked eastward and moved his lips. He hurried up a hill, squinted against the wind and watched clouds gather.

The clouds piled up like sooty cushions and the grinning herder waved them his way. He patted his dog and ran down the hill, scattering some sheep as he ran into their midst. The clouds followed.

The first rain came in stuttering spurts. Then the entire sky blackened and water broke forth in a joyous roar.

It fell on parched clover and thirsty flowers. It battered a skinny sapling and trickled down a great elm. It tumbled past a decaying fence into waiting streams. It washed dust off the

rocks and chased a spider into its hole. A squirrel zipped up a pine, and the sky clapped with glee.

The herder called to his horse on a near hill, but thunder drowned him out. Like a red-faced evangelist, heaven bellowed its message, and the message was hope.

After the rain the herder saddled his horse so he could check the pastures. Fresh, the gelding danced through chocolate puddles, sometimes stopping to taste his reflection. Birds perched on branches beside the gurgling streams and trilled a song—the herder thought it must be a hymn. He whistled too as he rode along.

Yesterday I rode along to deliver groceries and sheep salt to a herder who is packed out in Dead Man's Gulch.

It was slightly overcast for the first time in weeks and I was grateful for clouds to shield us in sparsely vegetated Bear Creek Canyon. Scattered jack pines shade the trail in only a few spots.

I followed Danny Vasquez, a camp tender, on his weekly trip to check on an old herder, John Jackson. I took my own horse, Lark, because I thought summer riding had conditioned her. She held up well for a young horse (being an Appaloosa helps), but she was noticeably wetter and more tired by day's end than the other two horses. Danny's palomino saddle horse, Blanco, is acclimated to mountain work and barely noticed the twenty-mile trip. The pack horse, a heavy-muscled brown mare, kept up a steady, ambitious pace all day.

We made good time. Danny set the pace, a trot, and we never walked a step all day. Trotting is my favorite gait, so I found it pleasant riding. At day's end I felt a little proud that I wasn't the least bit tired.

But Danny deflated me. He apologized for our slow speed and said it might have been more fun for me if he hadn't taken it easy for the horses' sake. My mouth must have dropped—not once all day did I have a sensation of dawdling.

Danny is about fifty, short, with straight black hair, stocky, and looks like Desi Arnaz in western disguise. He wears a tired and soiled brown felt hat that may have been a cowboy style when it was young—its crown and brim are disfigured with age.

His brown shirt and brown pants looked too heavy to me for a warm day, but then I suppose herders don't have great selection in their wardrobes.

I met him at 7 A.M. yesterday by the old trappers' cabin at the head of Bear Creek Canyon. When I arrived he was walking around the small building, feeling the durable old logs. He mounted his horse right away. The handle of a 30-30 peeked out of his saddle boot.

We set right off. At the trail's head Little Elk Mountain stood on the right and Commissary Ridge on the left. The trail wended beside Bear Creek, which normally roars through the canyon but this year, because of the drought, lapped quietly.

As we progressed, Big Elk rose up tall and grand on our right. The Fourth of July Ridge's sheer face was on our left. Serviceberry bushes and aspens already had turned yellow and weeds and infrequent oaks were red. They looked vivid against the brown earth and gray-blue rock ledges.

We came to Elkhorn Pass where someone once had mounted a massive set of elk horns in rocks just below the trail. Danny rode down into the rocks to point them out to me, but couldn't find them. They either recently had fallen or had been removed.

By nine o'clock the day was noticeably warm. We stopped where a stream splashed down tiered rocks to cross the trail. Danny dismounted and flattened himself on the ground, removed his hat, tilted back his head and caught the cold trickle. I've come to believe this is the best way to take water. I like the ritual even when I'm not thirsty. It offers a change in activity, a chance to stretch. And it's great to sprawl belly-down on the sun-warmed earth. When the water spilled into my mouth it splashed icy drops on my face and neck. The mountain water is heavy with minerals and satisfying and so cold it bites a dry throat.

At the brown cliff's base we passed a bubbling hot spring. Soon after, we turned into Dead Man's Gulch. Through the notch there, Caribou Mountain appeared.

They named the gulch, many decades ago, for a man who stayed in too late in the fall and apparently became trapped. A sheepherder found his dead body in the spring.

The vegetation changed in the gulch. Brome grass grew knee high to a horse and trees and bushes thrived. Wild raspberries wove a red pattern on the green fields. Forty-eight thousand sheep once grazed there, but regulations cut back on permits a number of years ago. Now eleven thousand head run there.

John Jackson, a large rugged man in his eighties, sat beside a log picnic-size table. His yellow Australian shepherd, Tuffy, lay by his feet.

John is six feet two, bald with gray tufts at his ears, blue-eyed and friendly. He originally is from Louisville, Kentucky, where he worked with Thoroughbreds. He has been in the West a long time and everyone says he's a good hand with horses.

The pack camp sits in a clearing at the base of a steep hill. Inside John's four-man tent were supplies and pack cupboard, but no bedroll. John has been sleeping in a teepee on the hill with the sheep because bears are a problem.

The pack camp resembles a national-park picnic site. Charred stones surround a fireplace with grill between the tent and table. We tied our horses at a log hitching rail and tossed our saddles over the log corral. Between two trees stretched a high log bench, a stationary hammock, perhaps. Upended runt logs are stools for sitting. I gave John a stupid compliment—I admired how natural everything looked.

John smokes a pipe continually. The rich smell of Prince Albert hovers around him. Most other herders I've met who smoke are cigarette men—some even roll their own.

Before this summer I wasn't aware that anyone rolled his own cigarettes anymore. When I watch men do this I feel I've been dropped out of the 1970s back onto the frontier. It's absorbing. First the man rustles in his pocket for cigarette papers. With some difficulty he separates one rectangle from the others. He draws a coated tongue up the paper's length. With thumb, forefinger and index finger, he forms a paper trough. Thump, thump, the tobacco spreads a spiny brown line as the man taps the can. His tongue wets the paper's edge and he seals the tobacco into a miniature tube. In symmetry and fullness the finished product is not much like mass-produced smokes, but hanging from blistered lips by a stubbly chin, it looks just right.

John insisted we stay for lunch. The meal he served lived up to his reputation as an excellent cook. Fresh sourdough biscuits, hot and bitey, juicy, mild chops from a freshly butchered lamb, and a macaroni and cheese side dish. Danny had told me macaroni and cheese was John's specialty and he hoped he'd serve some. But my tastes have been coddled by refrigeration—I found the cheese breathlessly strong. I admit the dish was unforgettable—hours later it remained on my tongue.

While John cooked he muttered to himself. His habit of talking to himself confused me at first. Sometimes I looked up to reply to something he said—then realized he had meant the remark only for himself.

One exchange with himself made me smile. While turning the chops he dropped his spatula. "Fall on the ground you son of a bitch!" he growled. "I was gonna wash ya anyway!"

John was talkative with Danny. They talked news—in town, a sheepherder had stepped in front of a car, somewhere a bear had caused a pile-up. Jackson himself had warded off a bear and he'd visited a neighborhood herder, Wild Dick Shoulders. Some of the talk was nostalgic—of men who used to herd on these hills, and outfits who no longer run sheep. Some was business, about lamb prices, grazing fees and availability of hay for winter.

When Danny and I got ready to leave, John saddled his horse to ride with us as far as the beaver ponds. His sheep were shaded up on the hill above his tent. He thought he'd try to get some golden trout for dinner.

About two miles from the pack camp he dismounted and tied his horse beside a serene pond. We watched fish milling as we rode by. John had hidden his willow pole and fishing line beside a pine tree. As we waved good-bye we heard him cursing because the beavers had carried away his pole, leaving a skinny track where they had dragged it up a hill.

Clouds moved in to cool the trip home. Yesterday's trip was among this year's prettiest, going over quiet terrain with good trail all the way. Only in one stretch did the path narrow, and footing was insecure. Horses' feet sank into soft earth and dislodged pebbles, sending the rocks rolling down the cliff's sheer face. Danny's horses kept up their usual trot. Lark would have hesitated to study things, but she didn't want to get behind so she trotted right over the uncertain footing.

Last week I caught myself complaining to a friend that work on the book had cheated me out of summer trail riding. I laughed when I realized what I'd said. The book research has brought me the prettiest rides I've ever been on.

My view of the back country has been unfairly beautiful—I've not had a glimpse of what the terrain is like in bad weather. I've faithfully packed my poncho for every ride, but it's stayed rolled up behind the saddle seat. If paths were rain-slick, or lightning crashed above ridge-top trails, riding there might not be so glorious.

Not all my trips have been relaxing. Some have been steep climbs, with heart-stopping scenery at the top. When I went with Ab Waters and Roy Cooper over Fourth of July Ridge it was vertical trail in many places. I walked that day so the men could pack Dolly because we were short one horse, and it was a grand place to hike, with top-of-the-world views.

We left the pack at Ab's camp and Ab rode back to Commissary Ridge with Roy and me. Now we had enough horses, but not enough saddles, so Ab and I rode the return trip in pack saddles. They felt a bit wooden, but with saddle blankets spread over the frame it wasn't too bad. I have a priceless mental picture of Ab riding along, like so much sheep salt, in a pack saddle. He rested his feet in the panniers and his long knees poked up near his chest. His hat brim drooped over his forehead to shield his face and a cigarette hung from the side of a grin.

On one precipitous hill I felt insecure riding in a pack saddle. The ground plunged vertically, and the horses rather skidded down on their fannies. I slipped off to walk, and gave my horse to Ab to lead. Ab, looking comical in his unusual seat, took the slide down unperturbed.

I had another uncomfortable moment, physically, not mentally. When the horses got a view of Commissary Ridge they started to run. Running in a pack saddle involves painful thumping against wooden crosspieces. I begged Dolly to stop, but she wouldn't. Fortunately, camp wasn't far away.

Descending from Commissary, Roy recalled the old days. We stood on a mountain and peered toward Brockman. "Can you picture," he said, "the road to Brockman, with fifty thousand sheep backed up on it waiting to be counted to go on the forest? And with them, horse teams pulling camps, and other teams pulling commissary wagons?" But that was the old days, of big outfits and big herds and big profits.

One day I rode with Juan while he moved sheep. Riding with a good herder is a tranquil experience. (Juan is a good herder who lets his amorous side distract him only in slack moments. When I trailed with him midsummer he rode beside me for a long time, trying, I thought, to demonstrate a whistle. After several miles I realized he was trying to convey the Basque word for—kiss.)

Unlike cattle drives, moving sheep is quiet. Cattle don't herd well and keeping them bunched involves racing and chasing on horseback, arm waving "Hyaaaaa's!" and other yells. But sheepherders disturb the flock as little as possible and rely on the sheep's own strong herding instinct and the old ewes, who know the way.

Perhaps, too, they are quiet in hopes of surprising wild creatures. When I see wildlife on these rides it makes my day. The herders, too, seem delighted, though for them it must be commonplace. The morning I rode with Juan I glanced over to see him waving excitedly. Grinning, he pointed above us where three elk were skylined on a ridge. He liked pointing out badgers and rabbits and tracks of big animals. Yesterday Danny seemed pleased when we surprised a moose in a beaver pond On our return trip the same pond was stirred up from a recent visitor, probably the same moose.

As we neared the clearing where I'd parked my trailer, it began to sprinkle. It was one of those dry western rains you can stand in and get only a speckled shirt and not the least bit wet.

It lasted ten minutes, barely long enough to chase away the flies and gnats. But just enough to leave behind a rainbow.

It hung above the dusty road and its foot poked into the canyon. I peered over the canyon rim as I drove by to see where the rainbow ended. Its pastel ribbons stopped atop a sagebrush plant. Too bad, the bush wasn't golden, only slightly yellow from early frost and lack of moisture.

When I visit the herders in the mountains I feel foolish about my earlier attempts to flush out religious attitudes. The stock questions I used to run through with every herder were: what happens to religious practice up here, does anyone maintain a prayer life, is religion discussed?

Almost every herder answered with a puzzled frown, a shuffled foot or an explicit head shake. No, was the answer, no religion anyone wanted to talk about.

Those questions, and answers, probably reflect our fragmented view of religion. To think a person who lived in that grandeur would need to set aside time to think about God, that someone who lives in such unspoiled beauty might need mechanical devotions, that theories or dogmas would be consequential up in the hills—those are a town dweller's ideas. He needs to devise moments of peace amidst his complicated life.

Not the herder. If he chooses, his whole life is a prayer.

XV

And I Will Dwell in the House of the Lord Forever

Manuel tightened the belly cinch against the buckskin mare, fashioned a leather knot to hold it and slipped the rear girth into its buckle. Wet grass, leather and horse smell teased his nostrils when he bent under the animal. He shut his eyes, straightened and gulped the excellent September air.

Its crispness bit his chest. It was good that two weeks ago he'd started moving the sheep back to lower ground. This morning he'd packed the pinto for a final trip; by late afternoon he would be at his sheep camp and gradually he and the flock would move down to the valley. Then, until snows were too deep, he would tend his sheep on the desert.

The mountains frowned with signs of an early winter. He might not beat the first storms. In August he'd seen trumpeter swans going south. Already, when quaking aspen stretched their arms in the morning wind, bleached leaves snapped and fell. Cold nights reddened the oak leaves. Insects slept in on cool mornings and bedded down early and fewer species joined hot afternoon choirs. Mosquitoes and horse flies had retreated in August and less bothersome hornets and yellow jackets took

over. Gnats by the millions thrived in the cooler weather and Manuel gazed out at the bedground through speckled air.

Weed patches sat silent without busy chiseler squirrels to skitter among them. In tight warm holes the puffed creatures lay in a sluggard's rest. A few thousand feet lower in the foothills, rattlers had discarded their hides and slithered away.

In the panniers were Manuel's teepee and grub and supplies. For three months the teepee had been home, wilting into a portable cushion whenever sheep *comida* and water required him to move. He wished someone could see how fine he'd whittled his aspen teepee poles, slender as a colt's legs. He hoisted them up and flung them behind dead logs.

He flung a canvas over the pinto's load. Over that he tossed his manila rope and secured it with a diamond hitch. He snapped another rope to the pinto's halter and led him over to the saddle mare.

When the three dogs, gnawing on a meaty sheep skull, saw Manuel mounted, they ran to his side. Brownie, the ablest, led off toward the sheep. It would be an easy day moving the sheep because the old ewes knew the way back.

Morning sunlight slapped the buckskin's neck when they reached the first ridge, and she broke into a peppy dance. Even the heavily loaded pinto's step quickened when light smacked his rump's shiny white patches. Then they were in the trees again.

Streaks from the mounting sun shattered on the pine crowns and fell in fragments on the forest floor. The horses picked their way over splotchy designs, and when the dew-slick path turned and dropped, the buckskin braced her front legs and skidded down. Her hindquarters hugged the earth. Manuel hoped the dogs were keeping the sheep together on the sheep driveway; he hollered to them. The sheeps' path was even steeper, too steep for the pack horse.

The pinto tested his front hoofs in several places while Manuel waited. The big gelding's bulky tan pack wavered to one side when he started down, and threatened to pull him over. But the horse recovered and got to work finding good footing on the next, steep, rockier stretch.

Many times Manuel had seen pack animals fall. Dead weight was harder for them to balance. Once he'd had an experienced mare tumble and roll over three times before falling into a wash four-men deep. She was unhurt, but it took him all day to recover his supplies.

In an hour, Manuel rode up to the Golden Gate, two large

spires which mark where the sheep trail turns. Moss gave the brown-gray rocks a golden glow and their name. The eastern sun, still kind, angled around the familiar rock pillars and love-patted Manuel's face and neck. A lone pillow cloud leaned on the spires, then bounced off to Alpine firs. The morning air, clean and thin, carried a mourning dove's poignant call.

A fullness in Manuel's throat and chest made him rein up. Three months ago he'd passed this landmark when the sheep started up to the high country. In the time since most mornings had been as this one, full of heartbreaking beauty. Nights had been painful at first, until he became friendly with them. Now the world below seemed uncertain.

The mare started up Fourth of July Ridge and Manuel's legs felt her shoulders tug. At the top he whoaed her though she wasn't a bit winded. He looked out.

Off to the east fifty miles, the Tetons' jagged spires reached for blue-purple sky. Below them, closer, Palisade Mountain's head showed bald where high altitude drew a neat line and forbade trees above it. Red Ridge, six miles to the east, blushed where red clay lay exposed, too high for timber cover.

The high, rough mountains of Wyoming's Gray's River Range pushed up against the eastern horizon.

South stood Big Elk and Little Elk mountains, molded in beautiful curves. A deep haze between the two peaks indicated Dead Man's Gulch. Manuel knew the canyon trail there. A deep V marked the trail head and the trail wended south along Wolverine Creek to Caribou Basin. Impressive Caribou Peak, whose gold brought fortune hunters in the 1890s, stood to the south and west.

In Gray's Lake Valley off west were the little towns Wayann and, a jump beyond, Henry, headquarters of some old-time sheep outfits and the hundred-year-old Henry store.

A hundred miles south the rugged Uinta Mountains of Utah looked like gentle blue mounds on the distant sky.

He looked north to Lightning Ridge. His boss had trailed sheep there before Forest Service regulations changed and he remembered the sheer, shaly footing which troubled even the best horses.

To the west was Wolverine Range and closer in was Skyline Ridge, cut by a brown strap, the weather-torn Skyline Road where stunning views rewarded Sunday drivers.

From his perch Manuel had a pretty view of Fall Creek and Fall Creek Basin's pines and greenery. He knew how good its grazing was for sheep. He glanced back at Golden Gate Pass and its varied greens.

He couldn't look out anymore. So he closed his eyes to look inside. From a far place in his memory a voice he didn't recognize read a Psalm verse, "This is the day which the Lord has made, rejoice and be glad in it." A bolder man might have lifted his head to the skies and called out the words. But Manuel only folded his leathery hands around the saddle horn and squeezed his eyes tighter.

Another voice called to him and he recognized it as his Indian grandmother's. When he was small she had taught him how Indians understood God as the Great Silence.

Sheep scattered in the brush below; he needed to leave. He flicked the reins on the mare's neck and she started down, cautious on the loose rocks.

Manuel moved on through silent trees.

What came to him next wasn't a prayer. It wasn't even a full-formed thought with sides and front and back and top.

It was just a feeling.

He dared not send it upward because the Lord might think it was a prayer and he was shy before the Lord. But the feeling swelled in his midsection and throat and would have ruptured his skin if he hadn't shared it.

So he laid it out like a clean picnic cloth on the forest where he rode. He let it flow out his pores to the firs and pines and white-trunked aspens. He sprayed it at sympathetic buckbrush and snowberry and wild daisies. He offered some to a bluebird that flew by.

The feeling was peace. Not peace as the world knows, but as the forest and mountains give.

For a moment his feeling almost became a prayer. Not a real prayer with start and pauses and "Amen" at the end. But he dared to let some of his precious feeling drift upward.

And if it had words the prayer would have said, "Let me keep this peace. When I'm back in town among other men, and women too, let me keep this peace."

Let me keep it. And share it.

Sheep have a strong homing instinct. When the herder turns the herd toward the valley, the ewes know the way back.

They recognize rich fields where ranchers put them to push for weight gain just prior to fall breeding. They remember bountiful fields where they've grazed, even years before, and will almost push down gates to get in.

Their memories are so excellent that lost sheep can find their way over fifty arduous mountain miles and arrive home months after the rest of the flock.

Herders like to be out of the high country before hunting season opens and before paralyzing winter storms strike. Then, too, alfalfa pastures in lower country are a favored spot for mating.

Vistas dobbed in Indian summer colors line the route home. The flats blossom with yellow rabbit-brush flowers. Maple bushes turn flaming red. Aspens are yellow, willows are orange and weeds are magenta.

Berries decorate bushes—white snowberries, red bear cherries and purple elderberries. Beside the path, purple ground grapes appear against shiny green-leafed plants.

Ewes arrive at rich pastures about a week before buck sheep do. The ranchers want their ewes gaining weight, but not overly fat, when mating begins. This practice, called flushing, results in better conception rates. Percentage of twins and triplets can range from 10 to 80, and good flushing is often the determinant.

Prior to mating season, bucks are pampered in an enclosed pasture and fed grain. When breeding begins, one buck will be expected to cover a hundred ewes. A most prolific animal, a buck sheep that accidentally gets in a band can impregnate seventy-five ewes before the herder discovers him the next day.

Ranchers truck the bucks to the ewes to conserve the animals' energy. But sheepmen recall when bucks trailed 150 miles to the ewes, were fed no extra rations and used at a ratio of one buck to 200 ewes. Still, conception rates were high.

In the pens, the rich feed and their instinctively high sex drive sometimes combine to produce homosexual behavior among the bucks. If one buck will tolerate it, the rest join in. The tendency disappears when bucks get with ewes in estrus.

Ewes come in heat every eighteen days. To observers, sheep mating seems perfunctory—the animals couple only briefly. But the herder who has a chance to observe the animals after the first rush of mating is over notices a subtle, though swift, courtship.

The buck approaches the ewe and strikes his front leg against hers. They both blat in deep voices heard only during mating season. The buck nuzzles the ewe's cheek. She coyly turns her face away while maneuvering her hindside toward him.

Sometimes the ewe turns down the buck and walks away. She then sidles up to another buck and preliminaries begin again. Courtship can last up to five minutes; mating lasts only seconds.

Outfits practice selective breeding to upgrade their herds. Good rams are an important part of this. Experts say that 80 to 90 per cent of gains in improving a trait like fleece weight comes from selection of good rams.

Most outfits use purebred bucks—white-faced for wool production, black-faced for good meat lambs. Cross-bred lambs can have good characteristics of both.

In choosing rams, ranchers look for animals that come from multiple births, made rapid weight gain as lambs and have good length. Other characteristics sheepmen look for are soundness of the mouth, straight legs, squareness, ruggedness, length in the hind saddle and a good head.

They select ewes for similar reasons—length, straight back line, substance, squareness on their legs and a refined, feminine head.

Suffolks are the popular black-faced breed. The Columbia is the most popular white-faced breed in range country because of its good mothering instincts, herd instincts and wool. The

Panama, developed by an Idaho breeder, is heavy-boned, substantial for range and also popular. A common breed, the Targhee, was developed at the U. S. Sheep Experiment Station in Dubois, Idaho, and is a good all-around animal. An experimental breed recently developed at the Idaho sheep station, the Polypay, was bred for increased profits through bigger lamb crops. A cross of Targhee, Dorset, Rambouillet and Finnsheep, the breed promises to produce twins twice yearly.

Before breeding season, herders and their bosses try to weed out older and barren ewes. They "mouth" and "bag" (check the condition of teeth and teats) to determine the animals' fitness to go on the range another year. A sheep's hardiness depends largely on the health of its teeth; if teeth aren't in good shape the animal can't do well in rough range conditions.

The crew examines milk bags for tears and disease. If a ewe doesn't pass both tests, the rancher sells her to a farm flock, where with easier conditions she can produce lambs for a couple more years. Ewes that consistently have failed to conceive will be sent to auction.

Unbred ewes are restless, dissatisfied animals, difficult to herd. Ewes in heat tend to congregate hopefully around any sheep that seems different, like a wether or a black. After breeding season the flock settles down and eats better. Though his band strays very little, the herder still must be watchful for predators, worry about weather and watch for disease.

Replacement ewes that joined the flock in the fall can be susceptible to a mysterious ailment called "big head." The malady, apparently caused by atmospheric conditions, causes fast, grostesque death. Newcomers are hit the hardest. Animals' heads swell with fluid. Some old-time herders bleed the animals by slashing their ears or even cheeks and tails to relieve the pressure, but there is little faith in that method generally.

In muddy conditions after thaws, the herder watches for hoof rot. Treatment involves trimming the hoof close and applying formaldehyde.

When ewes become heavy with lamb, the herder must be watchful for pregnancy disease. Pregnant ewes can tolerate little stress or shock, like interruption in feed and water. Overweight ewes are more susceptible. When the disease strikes, ewes act paralyzed, throw back their heads and die quickly. Dextrose administered immediately can save them, but it takes a trained eye to see the disease coming on.

About thirty days before lambing a crew arrives to tag the ewes. That involves shearing around the tail, genitals and milk bags. The practice makes for sanitary births, helps new lambs to find teats and eliminates manure-encrusted wool, which attracts flies. Flies that lay eggs in clusters of thousands can be a serious problem because when maggots hatch they bore into the body and cause death.

When snow covers the ground the herder begins to feed hay or hay pellets. In his winter pasture he and his animals aren't far from roads, cars and a town. Boundaries have fences.

The herder thinks about spring. There'll be tender young grass for his ewes. There'll be new lambs. And the cycle will start all over again.

I visit the mountains, and they are winter quiet.

In the valley, morning fog whitens the landscape and nighttime temperatures kick furnaces on, but middays are still warm for autumn activities.

Higher, layers of early snow crush roads and pastures, frozen weeds stand unnaturally erect and trees wear flocking. I lean on a deserted corral. Animal hairs that snagged on the wood during summer have collected frost, and dangle from poles like spaghetti. The earth creaks with the cold, insects are asleep or unborn and wildlife has waded to lower ground.

Nick and Emma Padilla have moved their camp down to Goshen, but even there the land is a smooth expanse of white. Like horses and unlike cattle, sheep paw through snow to get to grass so the ranchers don't have to feed hay yet. When that begins, it means several tons per day for each herd.

The coyotes followed the herds down, and Nick has been staying up late trying to protect the sheep. Sheepman Bill Siddoway told me he thought coyotes shouldn't be considered wildlife—they follow civilization like rats do. Coyotes are in every state now, and even New York sheepmen have problems with them, he said.

Inside Emma's camp, coal fuels the stove and it seems cozy.

(Partly from the heat, partly from Nick and Emma's hospitality.) When you're inside looking out the grasses seem sugar-coated and trees seem frosted, and it's a scene from fairyland. When you step outside, a raw wind blasts your fantasy.

Already Nick and Emma miss the mountains. Last summer Emma told me she sometimes hungers for female company when she's in the mountains. Now close to town she doesn't feel quite comfortable. She likes collecting her water in a bucket from the stream, walking alone through the trees and preparing fresh game for her stew pot.

The mountains in the distance do tantalize. When Jesus lived on earth He found refuge in the mountains. He chose His disciples there, gave the truth of the beatitudes and prayed. At the end, on Mount Olivet, He asked that His father "remove this cup from me." Maybe He too felt wistful about places He was leaving, the mountains where He walked and thought, the flowers and trees He wove parables about, the birds and blue skies and animals.

Sheep grazed the hillsides where He walked and He seemed to have a special affinity for them. Repeatedly He used the relationship of shepherd and flock to illustrate lessons. He said that one day, at the end, there would be one fold and one shepherd.

When He first came, shepherds were first to have news of His birth. When He comes back, a herder might again be first to know.

He might detect a quiet in the trees, a brilliance in the western sunset; he might observe odd behavior of ants and restless circling of birds.

He'd see if his ewes were nervous and hear a dog's inquisitive whimper and horse's snort. He'd notice a happy rustle of winds and weeds and flowers.

He might again be first to know because from his mountaintop that overlooks valleys and ranges he can observe so much.

But the herder's not waiting for Him to come back. All around him he feels His spirit, which never has left the mountains.